Garden and Landscape History

COTTAGE GARDENS AND GARDENERS IN THE EAST OF SCOTLAND 1750–1914

Garden and Landscape History
ISSN 1758–518X

General Editor
Tom Williamson

This exciting series offers a forum for the study of all aspects of the subject. It takes a deliberately inclusive approach, aiming to cover both the 'designed' landscape and the working, 'vernacular' countryside; topics embrace, but are not limited to, the history of gardens and related subjects, biographies of major designers, in-depth studies of key sites, and regional surveys.

Proposals or enquiries may be sent directly to the editor or the publisher at the addresses given below; all submissions will receive prompt and informed consideration.

Professor Tom Williamson, School of History, University of East Anglia, Norwich, Norfolk NR4 7TJ, UK.

Boydell & Brewer, PO Box 9, Woodbridge, Suffolk, England, IP12 3DF, UK.

Previous publications are listed at the back of this volume.

COTTAGE GARDENS AND GARDENERS IN THE EAST OF SCOTLAND 1750–1914

CATHERINE RICE

THE BOYDELL PRESS

First published 2021
The Boydell Press, Woodbridge

ISBN 978–1–78327–662–2

The Boydell Press is an imprint of Boydell & Brewer Ltd
PO Box 9, Woodbridge, Suffolk IP12 3DF, UK
and of Boydell & Brewer Inc.
668 Mt Hope Avenue, Rochester, NY 14620–2731, USA
website: www.boydellandbrewer.com

A CIP catalogue record for this book is available
from the British Library

This publication is printed on acid-free paper

Printed by TJ Books Limited, Padstow, Cornwall

To Bob, for constant support and encouragement,
and to the memory of my father,
who first got me to garden and do research

Map of Scotland by W. & A. K. Johnston, 1880.

CONTENTS

ILLUSTRATIONS

Tables

The author and publisher are grateful to all the institutions and individuals listed for permission to reproduce the materials in which they hold copyright. Every effort has been made to trace the copyright holders; apologies are offered for any omission, and the publisher will be pleased to add any necessary acknowledgement in subsequent editions.

ACKNOWLEDGEMENTS

I would like to thank above all Esther Read and Charlotte MacDonald for their unflinchingly thoughtful and tireless efforts to help me produce this book.

Professor Tom Devine, Professor Chris Whatley, Professor Phil Lyon, Catherine Smith and Liz Ashworth have generously given their time to read and respond to drafts of various chapters.

The period covered by the book ends well before the reach of living memories and most of the research was carried out in libraries and archives, local and national. I would like to thank Janet Carolan, the archivist of Dollar Academy, and the staff of Dundee Local History library, Perth and Kinross Archives, the Museum of Scottish Lighthouses at Fraserburgh, Fraserburgh Library, the National Records of Scotland Historical Research Room, the National Library of Scotland and the Royal Botanic Garden Edinburgh Library for their help and guidance.

The fun of research is not restricted to reading rooms, however, and any piece of research, however dusty, involves collaboration and sharing. I was privileged to meet Ian Tait of the Shetland Museum and Terrill Dobson of Angus Herbal Medicine Clinic, who shared their expertise on kale and herbs (respectively). Kenny Lauder of Rescuers Of Old Kilmadock (ROOK) took me on a fascinating tour of Deanston village. David Drummond not only filled me in about town dung but first put me onto the Internet Archive, the historical researcher's best friend, particularly during a pandemic.

Helen Hammond, Margaret Wallace and Aileen Elliott provided invaluable personal knowledge and family memories. Susanna Blackshaw and P. and G. Smith of Killearn Cottagers' Horticultural Society, Mima Magna of Markinch, Stuart Young of Kelso Horticultural Society, Douglas Lowe and Currie and District Local History Society, David Anderson, Kim and David Kirkaldy, Alistair Clark and Katherine Smith kindly allowed me to make use of documents and photographs from their family and society collections.

Karen Barton, Catherine Brown, Jeremy Burchardt, Christopher Dingwall, Derek Flynn, Ian Gasse, Sarah Hendry and Douglas Lockhart very kindly answered my questions on a variety of sometimes obscure issues.

Over such a long period of gestation, many others will have been contacted and consulted whose names have unfortunately escaped me: to them I also extend thanks, and apologies. It is also impossible to name all those – friends and strangers – who have spurred me on by telling me how interesting they found the topic of the history of cottage gardens. They know who they are and they have my thanks.

The patient, expert and heartening work of Caroline Palmer and Elizabeth McDonald of the Boydell Press was a vital part of the process of completing this book, and they too have my thanks.

Catherine Rice
Dundee 2021

ABBREVIATIONS

GM	*Gardener's Magazine*
GV	*General View of the Agriculture of the County of …*
HASS	Highland and Agricultural Society of Scotland
NBA	*North British Agriculturist*
NLB	Northern Lighthouse Board
NRS	National Records of Scotland
NSA	*New Statistical Account*
OSA	*Old Statistical Account*
RC	*Royal Commission*
SC	*Select Committee*
THASS	*Transactions of the Highland and Agricultural Society of Scotland*

COUNTIES OF THE EASTERN LOWLANDS BEFORE 1975, AND THEIR OLDER NAMES (IN USE UNTIL THE 1920S)

North-east
Nairn
Moray (formerly Elginshire)
Banffshire
Aberdeenshire
Kincardineshire (also known as the Mearns)
Central
Angus (formerly Forfarshire)
Perthshire
Fife
Clackmannanshire
Kinross-shire
South-east
West Lothian (formerly Linlithgowshire)
Midlothian (formerly Edinburghshire)
East Lothian (formerly Haddingtonshire)
Peebles-shire
Selkirkshire
Roxburghshire
Berwickshire

A NOTE ON OLD SCOTTISH WEIGHTS AND MEASURES

Area

Scots measurements of area remained in common use, on, for example, estate plans, until the early decades of the nineteenth century. They were replaced by imperial measurements, although 'falls' survived longer in some sources.

1 acre = 4 roods = 1.26 acres (6100 sq. yards) = 0.51 hectare
1 rood = 40 sq. falls = 1525 sq. yards = 1275 sq. metres
1 sq. fall = 38.1 sq. yards = 31.9 sq. metres

A fall could also be linear and many contemporary texts leave the reader to guess whether linear or square is intended.

Dry measures

1 boll = 0.6 gallons = 211.7 litres
1 peck = 0.9 gallons = 13.2 litres

(from the Scottish Archive Network *Weights and Measures Guide*, available at https://www.scan.org.uk/measures/capacity.asp)

Plate 1 Cottage at Burnside beside Currie Bridge, Midlothian, c. 1890–1900.

INTRODUCTION

The history of cottage gardens in Scotland is short, often depressing and sometimes puzzling. It is likely that cottage gardens in the English sense, with flowers and fruit as well as vegetables, were unknown in Scotland before the late 1700s. Throughout the eighteenth and nineteenth centuries travellers, agricultural and horticultural commentators and philanthropists bemoaned the cottagers' neglect more often than they noticed bright and well-kept gardens. A landowner lamenting the weed-infested and badly cultivated plots found them inexplicable in 'a country which supplies England with some of her best gardeners',[1] for this was the period that saw professional gardeners, often from humble backgrounds, prized for their skill, knowledge, education and reliability by the owners of mansions and palaces across Britain, Europe and even further afield.

While the gardens of the wealthy were, as innumerable studies have demonstrated, designed to display taste and culture, attract attention and admiration and provide delight, relaxation and fine produce in equal measure, the very different purpose and character of the cottage garden has been, until recently, largely ignored, or at best dealt with summarily. Even Edward Hyams – a notable exception to this approach – remarked that 'of course, it was in the great gardens of the rich and mighty, and not in the little gardens of the poor and humble, that both the art of gardening and the science of horticulture developed'.[2] Although he was in fact rescuing the cottage garden from obscurity, that 'of course' betrays a common perception of cottage gardens as also-rans, well behind the great gardens rather than in a different race.[3]

1 J. S. Menteath, 'On cottagers' garden societies', *Quarterly Journal of Agriculture*, vol. IV, 1832–4, 791–3, 791.

2 Edward Hyams, *English Cottage Gardens*, London 1970, 1.

3 Charles Quest-Ritson unashamedly focuses on the gardens of the 'middle and upper-classes, who are more literate and more conspicuous, and have left more traces of their passing' (Charles Quest-Ritson, *The English Garden: a social history*, London 2000, 2).

In Scotland the story of these gardens, even more than in England, is largely missing from both garden histories and social histories.[4] This book seeks to tell the story of this neglected subject. It is also an attempt to explain the significance of cottage gardens in the lives of the poor and their value to the workers, their employers and the wider rural economy and society.

Cottage gardens were important primarily as a source of food for workers' families. Among the poor, vegetables, albeit a very limited range for most, played a bigger part than in the diet of any other class. Farm servants, labourers and other workers based in the countryside, including handloom weavers, cotton-mill workers and miners, together with their families, might scarcely have survived without the produce grown on their plots. Nevertheless, the testimony of the workers themselves make it clear that they often also welcomed the solace a garden offered.

For landowners and employers – from farmers to coal-owners – the worker's garden was of importance too. Nearly all cottages, whether provided as part of the wage, as in the case of farm servants, some agricultural labourers, miners, railway workers and others, or rented to the casual labourers and rural artisans who were essential to farms and estates, came with gardens. They had an economic value first and foremost: they ensured the survival and health of workers and their families. Those who built, supplied, rented and (sometimes) maintained them also recognised, however, their social and moral purpose.

Vegetables were, in the eighteenth century, often shunned by the gentry. The opinion of E. H. M. Cox, author of one of the first histories of Scottish gardening, was that 'there is nothing in the history of Scottish gardening so puzzling as the relative importance of vegetables in the diet of the average Scot'.[5] And the rural poor who lived on little else apart from oatmeal grew only the basics, and often badly. 'The country-people are by no means aware of the advantages of a good and well-kept garden, nor of how much food for the family it will supply', complained the factor of an estate near Aberdeen.[6] A garden could be a burden to an overworked ploughman, yet – increasingly through the period covered by this book – there were many cottage gardeners who worked hard in their gardens, took pains to make them neat and tidy and attractive and took pleasure in them and in the crops they raised, including fruit and flowers. More than this: the profusion of cottage garden shows up and down the eastern Lowlands demonstrates the enthusiasm of many poor gardeners for growing an increasing variety of vegetables and fruit and both newly introduced and traditional flowers.

Cottage gardens were thus both functional and beautiful, both a larder and a pleasure-ground, both a curse and a delight.

---

4 Elizabeth Haldane's work is a rare exception (Elizabeth S. Haldane, *Scots Gardens in Old Times*, London 1934).

5 E. H. M. Cox, *A History of Gardening in Scotland*, London 1935, 118.

6 Alexander Thomson, 'On the settlement of crofters', *THASS*, 1837, 379–83, 381.

In England and in many other countries the significance of historic cottage gardens has begun to be recognised, albeit to a limited extent, in both publications and museum reconstructions.[7] Scottish cottage gardens of the past, on the other hand, have so far merited far less attention. With the single exception of the Weaver's Cottage at Kilbarchan, Renfrewshire, restored to its eighteenth-century state by the National Trust for Scotland,[8] there are no reconstructions of workers' gardens. The neglect in Scotland of this vital aspect of rural life is striking.

This book investigates cottage gardens and their gardeners during the period of greatest change in the countryside of the East of Scotland – from the time agricultural improvement (the 'Agricultural Revolution') took off in the form of enclosed fields, new crops and new farming techniques in the middle of the eighteenth century to the onset of rural depopulation in the decades before the First World War. It examines the crops gardeners grew (and those they didn't), the way they cared for them, what determined the size and location of the gardens, and gardeners' interest, or lack of it, in gardening. The wider picture is the economic purpose of these gardens; their role in workers' standards of living; their moral and aesthetic function; and the changing attitude of landowners and employers to them. As Kenneth Cox, E. H. M. Cox's grandson, has pointed out, 'Scotland's gardening history is intrinsically bound up with landownership and social class.'[9]

As well as recounting the untold story of cottage gardens in the East of Scotland, this book sets out to answer four questions. First, what did having a garden mean for the standard of living of rural workers and their families? Second, did their gardens have other meanings for them, unrelated to supplying food? Third, what were the expectations of landowners and employers with regard to the gardens and gardeners? And finally, what conditions had to be met to produce a well-tended plot?

Unlike the gardens of the well-to-do and the fabulously wealthy, fashion and taste played very little part in these plots. Poor gardeners were not immune from the passion for gardening that swept Britain in the nineteenth century, however, and there are glimpses of the influence of some of the changing fashions in plants and even garden layouts. Much more than changing fashion, though, it is the conditions in which these workers daily lived and worked, their relations with masters and landowners and the ties that bound them to their patches of land that truly explain the rural worker's garden.

---

7 For England Margaret Willes mentions recreated workers' gardens at Beamish Museum, Ironbridge Gorge Museum and the Weald and Downland Open Air Museum in Sussex (Willes, *Gardens of the British Working Class*, New Haven 2014, 2–3).

8 This well-built cottage belonged to a relatively well-to-do master weaver.

9 Kenneth Cox, *Scotland for Gardeners*, Edinburgh 2009, xxii.

This book is set in the Eastern Lowlands of Scotland, between the Moray Firth and the Tweed. From the temperate shores of the Moray Firth down the east coast to the rich arable lands of Angus and eastern Perthshire and south to the grain fields of the Lothians and the high pastures of the Borders, the Eastern Lowlands was the richest farming area of Scotland. The fertile band down the East Coast between the hill and mountain ranges and the North Sea was the most productive arable land not only in Scotland but at times in the whole of Britain. The thriving towns and seaports were supported by a hinterland of rich soils and sheltered carses (flat fertile land beside a river), and it was here that, from the middle of the eighteenth century, the landscape and its productivity were transformed by enlightened and progressive – and profit-hungry – landowners and a new breed of innovative tenant farmers. The new way of farming required increasingly large armies of full-time waged ploughmen, cattlemen, shepherds, day labourers and artisans such as wheelwrights and masons.

This countryside also had an industrial aspect. Thousands of domestic handloom weavers producing linen, cotton and wool cloth also lived in the countryside. Spinning mills and paper mills sprang up alongside fast-flowing rivers, at first water-driven and later powered by steam. In the later nineteenth century coal mining expanded rapidly, spawning colliery villages and 'rows' in Fife and the Lothians.[10]

The west of Scotland and the Highlands were very different in character from the Eastern Lowlands. The agriculture of the west and south-west counties was largely dairy, with much less arable. Farms were smaller and were usually worked by the family alone or with a few day labourers. Rural weavers here were mainly producing cotton cloth. Although coal mines cover the whole of the Central Belt, collieries in the west were newer, bigger and more profitable than in the East. Heavy industry grew around Glasgow, not in the east.

The Highlands and the western isles differed even more from the Lowlands in this period, and climate, land tenure, social relations and occupational patterns, not to mention language and culture, were so at variance with those to the South and East that comparison is largely irrelevant, as most social histories of Scotland have recognised.

THE APPEAL OF THE GRAND GARDENS

Garden historians have written much about the great gardens of the nobility and the wealthy in both England and Scotland, and about the innovative gardeners who worked for them. These are the great achievements, celebrated nationally:

[10] Coastal fishing communities, while economically significant and supplying the countryside with salt herring, make only fleeting appearances in this book. Fisherfolk tended to keep themselves apart from landward communities and their houses were so densely packed together along the shore that there was no space for a garden.

the heroic designs of Chatsworth and Blenheim, and the gardening kings, queens and generals – Joseph Paxton, William Robinson, Gertrude Jekyll – who shaped British gardens and created the English garden tradition. The tracing of this history feeds a huge public appetite for an understanding and appreciation of such gardens.[11]

Not so for cottage gardens, or indeed gardens of urban workers: given the popularity of garden history, and of gardens and gardening in Britain, astonishingly few books have been produced on the most common types. Perhaps the first modern account, Edward Hyams' *English Cottage Gardens*, is a model study, covering a thousand years and including both fascinating and highly knowledgeable discussion of the plants the cottage gardeners grew and the social context. More recently, Jenny Uglow's *A Little History of British Gardening*[12] set out to tell the stories of all kinds of gardens, both rich and poor – but has only one chapter on the gardens of the poor. Margaret Willes' comprehensive *Gardens of the British Working Class* covers the period from the seventeenth century to the present and every aspect of workers' involvement in gardening, from professional gardeners to Paisley silk weavers, and from the women who concocted herbal remedies to allotmenteers. By investigating not only the gardens but the men and women who worked in them, whether their own or their employers', Willes has shifted the ground of garden history, telling a story that challenges the often accepted view of worker gardeners' neglect, ignorance and conservatism. Willes also demonstrates how forms of landownership, legislation, philanthropy and socialist experiments produced very different types of rural garden.[13]

Jeremy Burchardt, in his equally pioneering *The Allotment Movement in England 1793–1873* (2002),[14] argues that allotments rented to rural workers in the poorest parts of England, who had only very small or no attached gardens, made significant and measurable contributions to living standards. They were large enough and productive enough to supply labourers with not only vegetables but wheat too, at around double the yields of farms. He also establishes that provision of allotments, initially a response by landowners to the threat posed by social unrest, later fulfilled a variety of objectives including behavioural change, by encouraging thrift, sobriety and industriousness for example. They held over

[11] See, for example, S. Festing, *Gertrude Jekyll*, London 1993 (and over twenty-one other works on her); Jane Brown, *The Pursuit of Paradise*, London 1999 (with one chapter on small gardens); Fiona Davison, *The Hidden Horticulturists: the untold story of the men who shaped Britain's gardens*, London 2019; Roderick Floud, *An Economic History of the English Garden*, London 2019 (one chapter, 'The People's Gardens', on urban and suburban gardens); J. Morgan and A. Richards, *A Paradise Out Of A Common Field: The Pleasures and Plenty of a Victorian Garden*, London 1990 (on walled kitchen gardens); Quest-Ritson, *English Garden* (nothing on working class gardens).

[12] Jennifer S. Uglow, *A Little History of British Gardening*, London 2004.

[13] Willes, *Gardens of the British Working Class*.

[14] Jeremy Burchardt, *The Allotment Movement in England, 1793–1873*, Woodbridge 2002.

the labourers the threat of eviction for misdemeanours and probably reduced rural crime, and were certainly believed to do so.

For Scotland there have been no full-length investigations of workers' gardens. Elizabeth S. Haldane's *Scots Gardens in Olden Times*, published in 1934, has a well-researched chapter on cottage gardens in the eighteenth and early nineteenth centuries. This and some references in E. H. M. Cox's *History of Gardening in Scotland* apart, scholarly effort has, as in England, focused principally on the spectacular grounds of castles, palaces and mansions.[15]

The lives and labour of rural workers are as much a concern of this book as their gardens. The searchlight turned on the lives of workers by E. P. Thompson's *The Making of the English Working Class*[16] inspired a generation of Scottish social historians, whose work has been central to this study. Thanks especially to the work of Christopher Smout, Tom Devine and Alexander Fenton, we now know a vast amount about the working conditions, standards of living, contractual arrangements, houses and material and intellectual culture of rural workers and how – and, more importantly, why – they changed in the eighteenth and nineteenth century. None of the major studies of groups of rural workers, however – Tom Devine's on farm servants and labourers, Ian Carter's on north-east farm servants, Norman Murray's on handloom weavers – mentions their gardens.[17]

Women played a crucial part in the economy of the countryside, particularly on the farm. As indoor servants and outdoor labourers on the farm they formed a larger proportion of the workforce than in England, and many of the handloom weavers were women too. Their role as hired workers has been researched by Devine and Iain MacDougall,[18] but their contribution to their own gardens (and indeed their domestic life) is still largely hidden from view, although they and their children were expected to join in the work on the garden. A recent study mentions women as weeders working on farms, in nurseries and in grand gardens,

15 Charles McKean, *The Scottish Chateau: the country house of Renaissance Scotland*, Stroud 2001; Marilyn Brown, *Scotland's Lost Gardens: from the Garden of Eden to the Stewart palaces*, Edinburgh 2012.

16 E. P. Thompson, *The Making of the English Working Class*, London 1963.

17 T. M. Devine, 'Social stability in the Eastern Lowlands of Scotland during the agricultural revolution 1780–1840', Lairds and Improvement in the Scotland of the Enlightenment, 9th Scottish Historical Conference, Edinburgh, September 1978; 'Scottish farm service in the agricultural revolution', in T. M. Devine (ed.), *Farm Servants and Labour in Lowland Scotland 1770–1914*, Edinburgh 1984; 'Social responses to agrarian "improvement": the Highland and Lowland clearances in Scotland, 1500–1850', in R. A. Houston and I. D. Whyte (eds), *Scottish Society 1500–1800*, Cambridge 1989; and *The Transformation of Rural Scotland: social change and the agrarian economy 1660–1815*, Edinburgh 1994; Ian Carter, *Farm Life in Northeast Scotland 1840–1914: the poor man's country*, Edinburgh 1979; Norman Murray, *The Scottish hand loom weavers, 1790–1850: a social history*, Edinburgh 1978.

18 T. M. Devine, 'Scottish farm labour in the era of agricultural depression 1875–1900' in T. M. Devine (ed.), *Farm Servants and Labour in Lowland Scotland 1770–1914*, Edinburgh 1984. Iain MacDougall, *Bondagers – Eight Scots Women Farm Workers*, East Linton 2000.

but not at all in the context of their own cottage gardens.[19] The evidence now makes it clear, however, that working- class women tended cottage gardens, and successfully, sometimes winning gardening competitions. The long tradition of women herbalists, it seems, also survived well into the nineteenth century.

SCOTTISH DISTINCTIVENESS

Why did so many Scottish cottage gardens so consistently fail to resemble their English counterparts? Margaret Willes contrasts the well-known chocolate box image of the English cottage garden – 'a charming, flower-filled space … before a picturesque cottage with thatched roof and roses climbing up the walls'[20] – with the much more complex and often less attractive reality. There was, however, some truth in this image, even if more cabbages were involved, for the southern half of England. William Cobbett, on a tour of the north of England, exclaimed 'Oh Lord! If the good people of this country could but see the endless strings of vine-covered cottages and flower gardens of the labourers of Kent, Sussex, Surrey and Hampshire.'[21] Towards the end of the century Flora Thompson depicts her Oxfordshire villagers unsentimentally in *Lark Rise to Candleford* and faithfully records the poverty and hardship of their lives. Nevertheless, 'they grew all the sweet old-fashioned cottage garden flowers, pinks and sweet williams and love-in-a-mist, wallflowers and forget-me-nots in spring and hollyhocks and Michaelmas daisies in autumn. Then there were lavender and sweetbrier bushes … almost every garden had its rose.'[22]

This kind of scene was not typical of the north of England, as Cobbett found, rather to his disgust, and the reasons for it are relevant to the Scottish case. The absence of flowers, trees, fruit bushes and herbs in cottage gardens north of the border (along with the general treelessness and poverty of the landscape), so frequently remarked on by eighteenth and early nineteenth commentators and travellers, can be largely but not wholly blamed on the climate. Today it is above all the climate that stands guard against the variety and lushness found in southern gardens. All the other obstacles faced by cottage gardeners described in this book bore, as we shall see, most heavily on the ornamental and decorative, as well as the perennial features of their plots. Poverty, insecurity of tenure and the customary movement of farm servants from farm to farm every few years, along with tradition and perhaps the influence of the historic inefficiency of Scottish agriculture, conspired, together with the scarcity of good examples of gardening, to keep most cottage gardens utilitarian – and very often unkempt

19 Deborah Reid, 'Unsung Heroines of Horticulture: Scottish Gardening Women, 1800 to 1930', PhD thesis, University of Edinburgh 2015.

20 Willes, *Gardens of the British Working Class*, 113.

21 William Cobbett, *Rural Rides* [1830], ed. George Woodcock, Harmondsworth 1967, 511.

22 Flora Thompson, *Lark Rise to Candleford* [1945], Harmondsworth 1973, 115.

and poorly cultivated. History, too, played a part. Edward Hyams suggests that the origins of the English cottage garden can be traced back to the flowers, herbs and fruit grown in medieval monastery gardens and then imitated by cottagers:

> a small and usually rectangular plot ... shaded by one or two apple trees ... [growing] a mixture of fragrant herbs and flowers taken from the wild, with currant and gooseberry bushes among them, and some wild strawberries .... In another part, ... the cottager grew cabbages, broad beans, leeks and onions.[23]

Monasteries in Scotland had important gardens and orchards too, but do not seem to have passed on their practice to post-Reformation gardeners to the extent that they did south of the border.[24] Nothing even approaching Hyams' vision was achieved by Scottish cottagers – as far as can be ascertained – before the late eighteenth century.

Later in the nineteenth century, gardens were on the whole smaller than their English counterparts and, for farm servants at least, were still needed mainly for vegetables. It was not until the second half of the century that flower classes overtook vegetable classes in village shows and, even then, where 'cottagers' classes' were included in the grander town and county horticultural shows they were invariably for vegetables only. While newspaper and magazine articles of the time urged cottagers to grow flowers, often stressing the moral as well as the aesthetic benefits, it is conceivable that flowers were still regarded by some influential lairds and clergy as too frivolous for the humble cottager.

RURAL WORKERS' GARDENS ACROSS THE GLOBE

Across northern Europe the 'vernacular garden' of the Middle Ages may have been very like Hyams' reconstruction: close to the house, fenced, with a limited mixture of vegetables, flowers, herbs and fruit, particularly the latter.[25] John Brinckerhoff Jackson argues that the expansion of farming at this time led to the demotion of the garden as the main provider of food but also a new role for it as a site of domestication of wild plants and of experimentation – 'a place devoted to study, small-scale beauty and perfection'.[26]

Studies of workers' gardens in this period have been carried out in several European countries, mostly in northern Europe. Allotments for urban workers were common on the edge of towns, and peasants in the countryside had larger

23 Hyams, *English Cottage Gardens*, 8.

24 Cox (*Gardening in Scotland*, 25) put down the absence of fruit from tenement gardens in towns in the fifteenth and sixteenth centuries to very short leases.

25 John Brinckerhoff Jackson, 'The past and present of the vernacular garden', in J. Dixon Hunt and J. Wolske-Bulmahn, *The Vernacular Garden: 14th Dumbarton Oaks Colloquium Series on the History of Landscape Architecture*, revised papers, Washington 1993, 12.

26 Ibid., 14.

plots. Large vegetable gardens (kålgårdar = cabbage yards) in eighteenth-century southern Sweden grew fruit, hops, hemp and tobacco, amongst other crops, enough not only to feed, clothe and entertain the family but also to provide a surplus for sale.[27] The landless in the Netherlands, Sweden and Flanders did not seem to have had subsistence gardens even in the seventeenth century and were entirely dependent on cash wages or money earned from selling handicrafts.[28] This seems also to have been the case in Scotland in the seventeenth century: labourers were paid in cash, farm servants had land.[29] Landlessness and proletarianisation amongst Scottish farm workers from the mid-eighteenth century and the related issue of the workers' demand for land is discussed in the present study.

In Ireland, in often similar circumstances and conditions to those in Scotland, both cottage gardens and allotments were encouraged by agriculturists, philanthropists and later government in much the same moralising vein as in Scotland. Like the Scots, Irish gardeners were berated for their lax ways and devotion to raising a very narrow range of vegetables.[30]

Research on workers' gardens across the world is still a sparsely cultivated field, but parallels and insights can be gleaned even from very different climates and settings. Lydia Pulsiphar's work on the gardens of slaves in the West Indies and the southern USA has shown that these were shaped by poverty, time constraints and their masters' views, just as Scottish cottage gardens were. On the island of Montserrat and possibly in other mountainous districts slaves could grow their own food if they had sufficient leisure time and space.[31] Some had Saturday afternoon and the whole of Sunday free and, on some plantations, once the daily task was completed, slaves could call their time their own.[32] The little yards were rarely ornamented with flowers and Westmacott suggests that the influence, example and practical help of the planter may have been key in the choice of crops – and whether or not to grow flowers. Westmacott's study of contemporary African-American gardens in the southern USA also draws attention to features that derive not from fashion but from necessity: little

27 Karin Hallgren, 'Vegetable growing by peasants in 18th century Sweden', Rural History Conference, Leuven, September 2017.

28 Richard Paping, 'The rise of landless households in the Dutch countryside ca 1600–1900'; Riika Miettinen, The livelihood strategies of the landless labourers in the early modern Swedish countryside'; Dieter Bruneel, 'Rural property offences and the changing nature of landless households in 19th century Flanders', all Rural History Conference, Leuven, September 2017.

29 T. C. Smout, *A History of the Scottish People 1560–1830*, London 1969, 134.

30 Jonathan Bell and Mervyn Watson, *Rooted in the Soil: A History of Cottage Gardens and Allotments in Ireland since 1750*, Dublin 2012.

31 Lydia Pulsipher cited in Richard Westmacott, *African-American Gardens and Yards in the Rural South*, Knoxville 1992, 13.

32 Ibid., 19.

evidence of thought-out design; climbers and cheap and reliable annuals rather than shrubs and perennials.

UNEARTHING THE HISTORY

Today, there are barely any traces left of the cottage gardens as they were planted and laid out by the rural workers. They have not survived or been reconstructed in the way that so many of those of the gentry and aristocracy have, nor is there anything like the documentary remains of individual grand gardens that historians can draw on, such as plans, invoices, detailed descriptions, photographs or sketches, that can be interrogated directly. Finding out what workers' gardens were like, what was grown, why, how and by whom, has mainly involved examining reports, books and articles that deal largely in generalisations and, of course, were written by the gardeners' social superiors. Writings by workers are few and far between.

Official and semi-official reports, products of the insatiable late eighteenth- and nineteenth-century appetite for data, have nevertheless been amongst the richest sources on the historical context and on the gardens themselves. Sir John Sinclair of Ulbster dominated the literature of horticultural and agricultural improvement[33] in the early nineteenth century and organised two enormous surveys, the *Statistical Account of Scotland* (henceforth the Old Statistical Account (*OSA*)), published in the 1790s, and the *General Views of the Agriculture* (*GVs*) of each county, published in the 1790s and early 1800s. Together, these accounts, while very uneven in their coverage, especially when it comes to the gardens of the poor, provide incomparable data on the changing rural scene. They are supplemented by Patrick Neill's much more critical *Report on the Orchards and Gardens of Scotland* of 1813. The *New Statistical Account of Scotland* (*NSA*), compiled in the 1830s and 1840s, though no more 'statistical' in the modern sense than its predecessor, was more consistent and again is a storehouse of information on a rapidly changing country.

Vast amounts of evidence on the lives of Scottish workers were amassed by Royal Commissions, Commissions of Inquiry, Select Committees and so forth on factories, mines, the employment of women and children, labour, housing and other issues. Nearly every one of these includes some information on cottage gardens gathered by the investigators, together with their subjective and frequently opinionated assessments. Just as importantly, these reports record the testimony of workers as well as that of their masters.

33 'Improvement' was the term used at the time, and later, to cover the changes in agriculture, industry and infrastructure which increased productivity and profits, and the social and intellectual developments that accompanied them. It is now generally regarded as too value-laden to appear without inverted commas, but is used in this book, without them, as a shorthand for those changes.

Many agriculturalists, clergymen, travellers and other commentators on the rural scene also described workers' gardens but, however honest and scrupulous, very few of them would ever have had much to do with such gardens, and not all would have examined them very carefully. The main purpose of much of this literature, too, was the organising of a more productive workforce, the raising of moral standards amongst the peasantry (often linked) or the upbraiding of a variety of neglectful groups – landowners, employers, clergy and cottagers themselves.

Agricultural journals, notably the *Transactions* of the Highland and Agricultural Society of Scotland (*THASS*) from the 1790s, illuminate the thinking of the landowning and farming classes, while gardening magazines, beginning with J. C. Loudon's *Gardener's Magazine* (*GM*) in the 1820s, and Scottish gardening manuals instruct on what could be grown and how crops should be raised. Local newspapers and their reports of cottage garden society flower shows have allowed drilling down to the village level and have even, with the help of census records, yielded the names and occupations of working-class gardeners.

The rarest but in many ways the most valuable sources are the memoirs, poetry and essays by rural workers. Their writings, together with biographies of a few of these men, well known in their times but now forgotten, have also allowed glimpses of their gardens and insights into their relationships with them. The contrarian journalist Alexander Somerville, the poets Alexander and John Bethune, James Thomson of Kenleith, John Nevay, Alexander Laing and others, ploughmen pamphleteers and the self-taught botanists George Don and John Duncan, all make their lively appearances in this book. Many of them also had their agendas, whether it was agitating for better conditions for farm servants, exhorting their fellows to help themselves or recalling a golden past era.

Attitudes to and preoccupations with rural life show up in paintings and in the picture postcards of the day. A taste for paintings of rural life arose in the second half of the nineteenth century, when it had already started its long decline. Cottages were more often the subject than the gardens, often shown with their elderly inhabitants at the door, emphasising the mournful and nostalgic atmosphere. Two later well-known exceptions are Arthur Melville's *A Cabbage Garden* (1877) and James Guthrie's *A Hind's Daughter* (1883).[34] Both depict strong, independent figures dominating their brassica-filled plots. Picture postcards, by contrast, showed pretty thatched cottages and village streets in their hundreds, although few images led past the front fence or into the back garden, which is where we would like to be taken.

This is a first attempt at a history of cottage gardens in Scotland and as such is exploratory, tentative and limited in time and in place. The research carried out for this book is bound to have missed both much documentary evidence and many people with personal knowledge and memories passed on by grandparents of gardens long ago. To a greater extent than is desirable, the investigation has

[34] Both paintings are in the collection of the National Galleries of Scotland. See www.artuk.org.

been led by the evidence. The lighthouse keepers, for example, a tiny group, feature rather larger than one might expect because so much of their working lives has been documented and preserved, including their gardens and gardening practice, while railway workers' gardens have not been tracked down at all. I hope that there will be readers who will be driven by these gaps and unevennesses to investigate the gardens of their own localities, redress balances and deepen and complete the picture.

Each chapter of this book picks up its subject early on in the period and follows it through to the First World War. Chapter 1 gives a birds-eye view of the Eastern Lowland landscape as it underwent radical change over the period. Chapters 2 and 3 deal with two types of garden: the basic kailyards of the farm servants and the cottage gardens of labourers, artisans and others. Two short chapters, on potato grounds and on middens, discuss these fundamentals of cottage gardening. A chapter on the rural diet the gardens helped to supply follows. Chapter 7, on flower shows and well-kept garden competitions, recounts the enthusiasm of many rural workers for gardening, encouraged by paternalistic landowners and employers. How rural workers learned to garden and found out about new plants, and how a few found gardening a stepping stone to scientific inquiry, is dealt with in Chapter 8. The final chapter looks at how different social classes viewed cottage gardens, both as an ideal and as a tool of social engineering.

For most cottage gardeners gardening was just another chore – an everyday and unpoetic activity that, like cooking or housework, was essential to the survival of the family but that did not merit much reflection. But Peter Still's idyllic vision of the garden, distilled from a life of pain and misery, was shared by many other worker poets at this time. Still worked as an agricultural labourer in Aberdeenshire and wrote his long poem, *The Cottar's Sunday*, in tribute to Burns' *The Cottar's Saturday Night*. His model cottager spends his Sunday surrounded by his model family, reading the Bible and attending church. First, however, he inspects his crops of oats and barley and his kailyard:

> Oh how deligthfu' dawns that blissfu' morn,
> When nature wears her loveliest robes o' green!
> When fairest flow'rets ilka field adorn
> An joyous June leuks laughin' o'er the scene!
> The cottar frae his ha' comes forth alane,
> An' doun his rigs or kailyard saunters slow;
> Wi' thoughts contemplative, wi' soul serene,
> He marks the dewy daisies round him blow,
> While, borne on wings of love, his feelings heavenward flow.[35]
>
> The Cottar's Sunday

[35] Peter Still, in *The Cottar's Sunday and Other Poems, chiefly in the Scottish dialect*, Aberdeen 1844.

It is important to bear in mind through the account that follows, which focuses mainly, and inevitably, on the practicalities and privations of cottage gardening in the harsh climatic, social and economic conditions of Scotland, that the rural workers' gardens could also inspire and exalt. But for most of these workers Peter Still's vision and sentiments would have rung hollow. For them the meaning of the garden was a utilitarian and functional one, as attested by the frequently expressed dismay of those who might have described themselves as the guardians of rural society – the clergy and the landowners.

CHAPTER 1

THE CHANGING LANDSCAPE

The countryside in which the workers laboured and cultivated their gardens changed out of all recognition between the middle of the eighteenth century and the start of the twentieth. Increasingly intensive and scientific farming transformed the landscape, not only bringing in unprecedented profits but also depriving most of the people of the tiny patches of land from which for centuries they had scratched a semi-independent living and marshalling them into disciplined, closely supervised armies of wage-earners. T. M. Devine has characterised the break with the past in Lowland Scotland in the later eighteenth century as 'more decisive and [the] social transformation therefore more disruptive' than in England or even the Highlands.[1] This transformation did not only affect agriculture. Industries, mines and railways also took over the countryside and demanded equally large, closely supervised workforces.

It was no accident that the Eastern Lowlands became the most prosperous and innovative agricultural area of Scotland – the envy of Europe by 1830, in T.C. Smout's words.[2] This gently undulating landscape – a narrow coastal strip in the north widening to cover half the country further south – is sheltered from the north and west winds by a series of mountain and hill ranges: the Cairngorms and Grampians in the north, the Sidlaws and the Ochils in lowland Angus and Perthshire and the Pentland, Lammermuir and Border hills in the south-east. These ranges help to make for a relatively dry and temperate climate. In the Carse of Gowrie, between Dundee and Perth, the Sidlaws protect the most favoured stretch of farming land in the country from the cold dry easterly winds. 'There is no country in Scotland, which enjoys a climate so mild and favourable to vegetables as this'.[3] The districts along the Moray Firth in the north, between Stonehaven and Dundee, the Forth basin and the south-east have the best soil for arable farming: deep rich loam and workable clay.

In spite of its advantages, the latitude of the Eastern Lowlands keeps it a good deal chillier than England – two degrees cooler than the south of England on

1 Devine, *The Scottish Clearances: a history of the dispossessed*, London 2018, 117.

2 Smout, *Scottish People*, 291.

3 James Donaldson, *GV Carse of Gowrie*, Edinburgh 1794, 7–8.

average. During the eighteenth and a large part of the nineteenth centuries the Little Ice Age still held Scotland, more so than England, in its grip, although it was the north and west of Scotland that felt it the most. Warming began around 1700 but there were still years and indeed whole decades of icy winters and cool wet summers, as well as storms, floods, droughts and blizzards. These were years of scarcity and high grain prices, when the harvest was not finally gathered in before mid-winter.[4] This weather pattern is likely to have played its part in the changing landscape: lairds and farmers may have been encouraged to experiment with new crops and techniques by warmer weather in the 1740s and 1750s and again in the 1790s and 1800s. The 1820s and 1830s also saw higher temperatures.[5] Colder weather returned in the 1880s – the beginning of the agricultural depression, the peak years of emigration to North America and the flight of farm workers from the countryside.[6]

THE IMPACT OF IMPROVEMENT

Until the second half of the eighteenth century most of the cultivated Lowland landscape was still patterned with the strips (rigs) of the fermtouns, the settlements of tenant farmers and labourers. It was not uncommon for twenty or more tenants, sub-tenants and cottars to inhabit just one fermtoun, although single tenancies were also already well-established in many places. Much of the farming was at a subsistence level. The arable portion of the farm was made up of unenclosed ridged strips, known as runrig, covering the continuously cultivated infield and the less fertile outfield, with each tenant leasing rigs in both for growing oats, barley and peas and beans. Rough grazing was held in common and 'waste' ground supplied fuel and building materials. Each household also had a fenced yard for vegetables, known as the kailyard.

Villages, with curling, irregular streets and odd-shaped market places, were rare at this time outwith the south-eastern counties of Berwickshire and East Lothian (Plate 2). Fermtouns were common and universal throughout the Lowlands. 'The contrast between the well-developed English farm village and the Scottish fermtoun reflects the very different rural traditions within the two countries.'[7] The few tradesmen needed by the farmers lived on the fermtoun, combining wright work, smithing, shoemaking or handloom weaving with subsistence farming.

There were two distinct phases to the transformation of the countryside. The first phase began in earnest around 1760, with the rapid adoption of improved farming techniques and enclosures, the building of new roads and farm steadings

---

4 Graeme Whittington, 'The Little Ice Age and Scotland's Weather', *Scottish Geographical Magazine*, vol. 101, no. 3, 1985, 174–8, 175.

5 H. H. Lamb, *Climate, History and the Modern World*, London 1995, 244, 248, 251.

6 Ibid., 254.

7 R. N. Millman, *The Making of the Scottish Landscape*, London 1975, 153.

Plate 2 The village of Garvald, East Lothian, no date.

(farmyards and buildings) and the planting of forests and shelter-belts. The 1780s saw the erection of rural water-powered textile and other mills. By 1830 most of Eastern Lowland agriculture and the landscape had started to develop their modern contours. In the second phase, from the 1850s, rising profits enabled farming to become highly capitalised. It was at this time that the railway network expanded and many new and larger mines were sunk. And then, within a few decades, began the gradual depopulation of the countryside.

Regional contrasts were sharp, and remained so. The Lothians were in the vanguard of agricultural progress throughout the period, with the largest and most profitable arable farms, growing mainly cereals. The Lothians were also where the fastest-growing coal mines were. The north-east became the main cattle-rearing region, while on the marginal land small farmers struggled to convert moorland ('waste') into crop-bearing soil; agriculture, fishing, forestry and quarrying accounted for nearly all rural employment. In the central counties of Angus, Perth and Fife there was mixed arable and stock-rearing farming, with mining in the south of Fife.

Before the 1760s energetic and visionary landowners such as John Cockburn of Ormiston in East Lothian were enclosing rigs in fields, rebuilding farm steadings and giving their most enterprising tenants long leases conditional on their committing to new ways of farming. After the 1760s the trickle of experiments became a flood. Aristocrats such as the dukes of Perth and Gordon, the earls of Haddington and Hopetoun, the law lords Gardenstone, Monboddo and Kames and many lesser landowners took up the new fashion enthusiastically, although few of them farmed themselves. After several fortunes had been lost in these pioneering years, revenues and rents started to rise from the 1780s – and it was the educated, energetic and risk-taking factors and other professionals who, together with the large tenants, undertook improved regimes and made farming commercial and profitable.[8]

THE TRAVELLERS' VIEW

Eighteenth-century travellers to Scotland, however, curious about the new 'North Britain', were still finding poverty at all levels of society and glaring contrasts between the old and the new. The Welsh naturalist Thomas Pennant, visiting Scotland in 1769, described regular fenced fields with good crops surrounding new stone farmhouses throughout the Lowlands, but 'a mixture of the old negligence [was] left amidst the recent improvements, which look like the works of a new colony in a wretched impoverished country'.[9]

A young French aristocrat and his tutor touring Britain in 1786 were unimpressed by the poor soil, lack of trees, sparse population and long swathes of uncultivated land as they crossed the border at Berwick-on-Tweed.[10] Nearer Edinburgh they came across evidence of better farming but, after taking the ferry across the Forth, they decided Fife was 'scarcely better than desert, with farms very few, badly built, poor, cottages rather than farms … '.[11] Around Perth, wooded hills, excellent soil and farming methods and flourishing cottagers' plots struck them favourably again but between Dundee and Aberdeen the prospect was grimmer. Plantations along the Moray Firth were a welcome sight, but the farms in that region were covered in white sand[12] and dispiritingly poor.[13] Nowhere in Scotland did they come across countryside such as that around Ripon in Yorkshire, which they had described as 'like a garden'.[14]

These travellers all arrived fully fitted out with the unavoidable prejudices and expectations that accompany those venturing into unknown lands. Dr Johnson famously observed, of Kincardineshire, 'that it was wonderful to see a country so divested, so denuded of trees',[15] although large-scale tree planting by the duke

8 Smout, *Scottish People*, 292–9; Devine, *Transformation of Rural Scotland*, 70.

9 Thomas Pennant, *A Tour in Scotland 1769* [1776], Perth 1979, 42.

10 Norman Scarfe (ed.), *To the Highlands in 1786: the inquisitive journey of a young French aristocrat*, Woodbridge 2001.

11 Ibid., 128.

12 A huge sandstorm in 1694 covered the land near Nairn, destroying fertile farmland and livelihoods.

13 Ibid., 161.

14 Ibid., 70.

15 James Boswell, *The Journal of a Tour to the Hebrides with Samuel Johnson* [1785], London n.d., 65.

of Atholl and many other lairds was already under way. Nevertheless, the overall impression of poverty and backwardness compared to England was undeniable.

By the first decades of the nineteenth century the planting of shelter-belts was moderating the destructive and chilling winds. Draining marshland made previously unusable land available for cultivation and was also said to have reduced the incidence of fogs and hoar frosts.[16] Hedges, fences and drystone dykes marched across the land protecting the carefully tended fields. Manuring and liming raised the fertility of the soil, including that of the former 'waste' land – moorland and hill pasture. Rotations including turnips and sown grasses further improved fertility and livestock rearing.

A more cheerful picture greeted the poet laureate Robert Southey in 1819, when he toured Scotland in the company of Thomas Telford. His account of the countryside between Dundee and Arbroath (about fourteen miles apart) was far more complimentary. Where the two Frenchmen in 1786 had found in this district 'some of the least cultivated lands'[17] of their whole trip, Southey praised 'a highly cultivated country the whole way. Large fields and a beautiful harvest, such as to make the heart glad … . Many neat cottages and houses of a better description.' This prospect was nevertheless marred 'as usual' by 'filthy women with their hair in papers'. Southey noticed both the surprising number of women working in the fields and the equally startling quantity of prostitutes in Arbroath.[18]

NEW PATTERNS OF RURAL SETTLEMENT

William Cobbett, visiting in 1832, was astounded equally by the productivity of Lothian farming and by the absence of workers and their dwellings.[19] But the Lowland clearances of the cottars, while every bit as effective as that of their Highland counterparts, passed almost without notice. Evictions appear to have been carried out without protest. There was alternative and often better-paid employment in nearby towns and cities. Emigration to England and to North America was becoming an option for some, but overseas emigration from the Lowlands did not reach the levels of Highland emigration until the

16 '… these dingy mists have disappeared before the enterprising agriculturist. Thorough drainage and improved cultivation have completely revolutionised the county of Fife – have changed it from an unhealthy swampy waste … into a rich agricultural county' (James MacDonald, 'On the agriculture of the county of Fife', *THASS*, 1876, 8–9). The impact of marshland draining on climate is now well established by modern research (see N. Schneider and W. Eugster, 'Climatic impacts of historical wetland drainage in Switzerland', *Climatic Change*, 80, 2007, 301–21).

17 Scarfe, *To the Highlands*, 140.

18 Robert Southey, *Journal of a Tour in Scotland in 1819*, ed. C. H. Herford, Edinburgh 1929, 58–60.

19 William Cobbett, *Tour in Scotland and in the Four Northern Counties of England in the Autumn of the Year 1832*, London 1832–3, 89.

end of the period. In the north-east, however, alongside the amalgamations of small farms into larger units, croft or smallholding numbers increased, as landowners realised the cheapest and most efficient way to bring moorland into cultivation was to lease parcels to cottars or small tenants at minimum rents, or even none.[20]

In fact the Lowland clearances were so thorough-going and successful that many landowners, alarmed at the evaporation of a pool of casual and seasonal labour and the flight of tradesmen to the towns, as well as needing local markets for their crops, began to create whole new villages to hold them in the countryside. This phenomenon – planned villages – was mainly to be found in the north-east, but there were many further south too.[21]

The new villages, or in some cases additions to existing villages, were often laid out on uncultivated land offered rent-free at first to attract tradesmen and 'manufacturers' (weavers), who would bring it into cultivation and be available for the harvest work. Each settler had a piece of land for a house and garden, as well as a few acres for raising oats and flax or grazing a cow or two. One of the earliest and most successful was Laurencekirk in Kincardineshire, the ewe-lamb of the landowner, Lord Gardenstone – 'he is exceedingly fond [of it], and has written a pamphlet upon it, as though he had founded Thebes' wrote James Boswell, who, unlike most visitors, was not particularly impressed by the village.[22] The fashion for planned villages flourished, even if not all the villages did. By 1850 there were nearly 500 of them in all, instantly recognisable today with their broad straight streets lined with solid stone-built houses and large gardens behind.[23]

Informal settlements of labourers and artisans were allowed to spring up on marginal land, sometimes rent-free in return for farm labour. In addition, a new pattern started to emerge of farm workers' and labourers' cottages with attached gardens, some new, some repurposed farmhouses, some still the old cottars' hovels, sited on farms and not in villages. Competition for labour with the new industries, both rural and urban, and the need for year-round skilled workers on mixed arable and livestock farms meant that landowners and farmers were obliged to continue to house the ploughmen, shepherds and cattlemen – and often labourers too – as they had done in the past. In fact, according to Devine, 'the labour-intensive operations of improved mixed husbandry actually demanded more workers per acre, especially of dependent women and children, than the old system.'[24] Labour shortages were more common than in the south

20 Devine, *Scottish Clearances*, 161.

21 D. G. Lockhart, *Scottish Planned Villages*, Edinburgh 2012, 1.

22 Boswell, *Journal of a Tour to the Hebrides*, 65.

23 Lockhart, 'Planned villages in north east Scotland, 1750–1860', in J. Frew and D. Jones (eds), *The New Town Phenomenon: The Second Generation*, St Andrews Studies in the History of Scottish Architecture and Design IV, St Andrews 2000, 25.

24 Devine, 'Social responses to agrarian "improvement"', 164.

of England, where specialisation in cereals, requiring quantities of seasonal labour but not year-round work, and the lack of alternative jobs in industry led to unemployment, underemployment and falling real wages.[25]

HIGH FARMING AND DEPRESSION

Scottish agriculture boomed in the period characterised by the farming methods collectively known as 'high farming', the 1850s and 1860s. Greater capital investment and technical advances such as artificial fertiliser and labour-saving machinery increased arable acreage, yields and profitability and also encouraged larger farms, with larger workforces. Increased revenues allowed the erection of imposing new farmhouses, more like mansions, and large and often elaborately decorated farm steadings with covered cattle-courts and stabling for fourteen pairs of horses or more. Farm servants and labourers eventually benefited too – on the whole – as their wages rose and better cottages were built. It was in these decades, too, that railway lines started to criss-cross Scotland, while quarries pitted the land, scarred in Fife and the Lothians by ever larger coalfields. Mining villages sprang up or expanded, some to become small towns.

The inevitable result of more mechanisation on the farms was fewer jobs. The number of man-days a year needed to work on an acre of barley fell from twenty-two in 1840 to twelve in 1914.[26] The arrival of cheap wheat from North America hit Scotland, with its largely mixed farming, less hard than England, but had an impact nonetheless. Over the whole country, the percentage of employed men in agriculture fell from 30 per cent in 1851 to 14 per cent in 1901.[27] Farms shed casual labour. Ploughmen and other skilled workers, with little prospect of advancement or a chance to acquire even a small farm, emigrated abroad or left for more congenial or better-paid work in towns and mines. Hamlets and small villages shrank or disappeared; farmers pulled down cottages in efforts to keep paupers away and poor rates down. At the same time, 'rural districts are becoming every day more agricultural' wrote the prominent farmer Robert Scot Skirving in 1862: weavers were turning to labouring and tradesmen retreating to towns and villages.[28]

The 1891 census recorded that, for the first time, over half the population of Scotland was living in urban centres, although the Borders and the north-east remained largely rural until the Second World War.[29] Those who remained on the land mostly gained: real wages rose, although they still lagged behind those of other occupations. And as farmers and landowners saw their ploughmen heading

---

25 Devine, 'Scottish farm service', 2–3; Burchardt, *Allotment Movement*, 9–10.

26 Smout, *Scottish People*, 61.

27 Ibid., 58.

28 Robert Scot Skirving, *Landlords and Labourers*, Edinburgh 1862, 11.

29 Michael Flinn (ed.), *Scottish Population History*, Cambridge 1977, 313.

off they began to commit to serious cottage-building programmes, providing (outdoor) toilets and running water in order to compete with the relatively more salubrious conditions of the towns. Rural slums still pockmarked the countryside, especially in the mining districts but also in many villages. These demographic and economic changes had a profound impact on rural society. At the same time, the charms of rural life were drawing the middle classes to the countryside, where they built spacious, well-appointed and ornamental cottages with elegant front gardens and began to change the character of many villages.

THE TRANSFORMATION OF AN ANGUS PARISH

Maps of the parish of Monikie, near Dundee, clearly show the kind of changing landscape the travellers wrote about. William Roy's military map of the mid-eighteenth century (Plate 3) shows the carefully laid out wooded and enclosed policies of the earl of Panmure, a scattering of fermtouns and their rigs – and a single road. Most of the land was rough grazing and waste – bog or exposed, rock-strewn moorland. Less than fifty years later (Plate 4) Ainslie's map shows roads, plantations and fewer fermtouns. Cottertouns were still scattered about, now housing farm servants and labourers. Rents in the parish had doubled by the 1790s, small multiple tenancies had been 'engrossed' into single tenancies to make bigger farms, smart new farmhouses and steadings were being built, turnip and 'artificial grasses' such as clover were being sown, ten-strong ox teams were replaced by pairs of horses at the plough, while selective breeding, young plantations and the cultivation of former waste land were all galloping apace.[30] Ploughmen's wages had tripled in thirty-five years and even women labourers' had doubled.[31] A new turnpike road between Dundee and Arbroath was being built. The minister of Monikie also noted the falling population of his parish, caused by the amalgamation of farms and the 'extermination' of many cottagers.[32]

The first edition of the Ordnance Survey map of Monikie, surveyed in 1858 (Plate 5), shows change continuing. There are more roads (and railways nearby), meaning greater mobility even for the poor. The cottertouns and their inhabitants, including most of the handloom weavers, have vanished, but there are a now couple of small villages. By 1900 (Plate 6) the uncultivated moorland had been replaced by a plantation and there were fewer but larger steadings. Hillsides were being excavated for stone and the villages had expanded: the structure of today's landscape has emerged.

[30] *OSA* Monikie, Angus, vol. IV, 344.

[31] Ibid., 347.

[32] Ibid., 345.

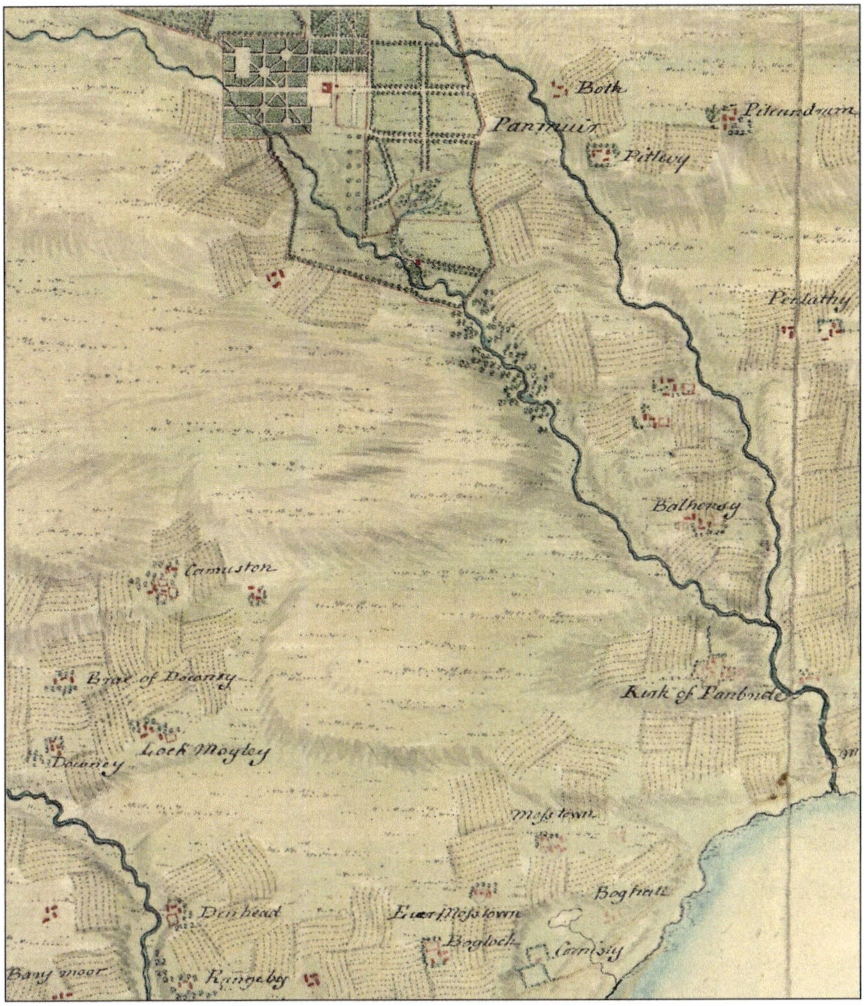

Plate 3 Monikie, Angus in the middle of the eighteenth century. Detail of Roy Military Survey of Scotland, 1747–52. (Shelfmark CC.5.a.441 19/4B)

GARDENS OF THE WELL-TO-DO

The east of Scotland is dotted with the impressive and elegant homes of the aristocracy and gentry: the eighteenth-century palaces and mansions such as Floors, Mellerstain, Paxton, Hopetoun and Dalkeith south of Edinburgh; Scone Palace, Blair Drummond and Blair Atholl in Perthshire; and the castles of Fife, Deeside and Moray. Gardens, designed landscapes and wooded policies added to the impact they made on the surrounding country.

Plate 4 Monikie in the 1790s. Detail of Map of the County of Forfar or Shire of Angus by John Ainslie, 1794.

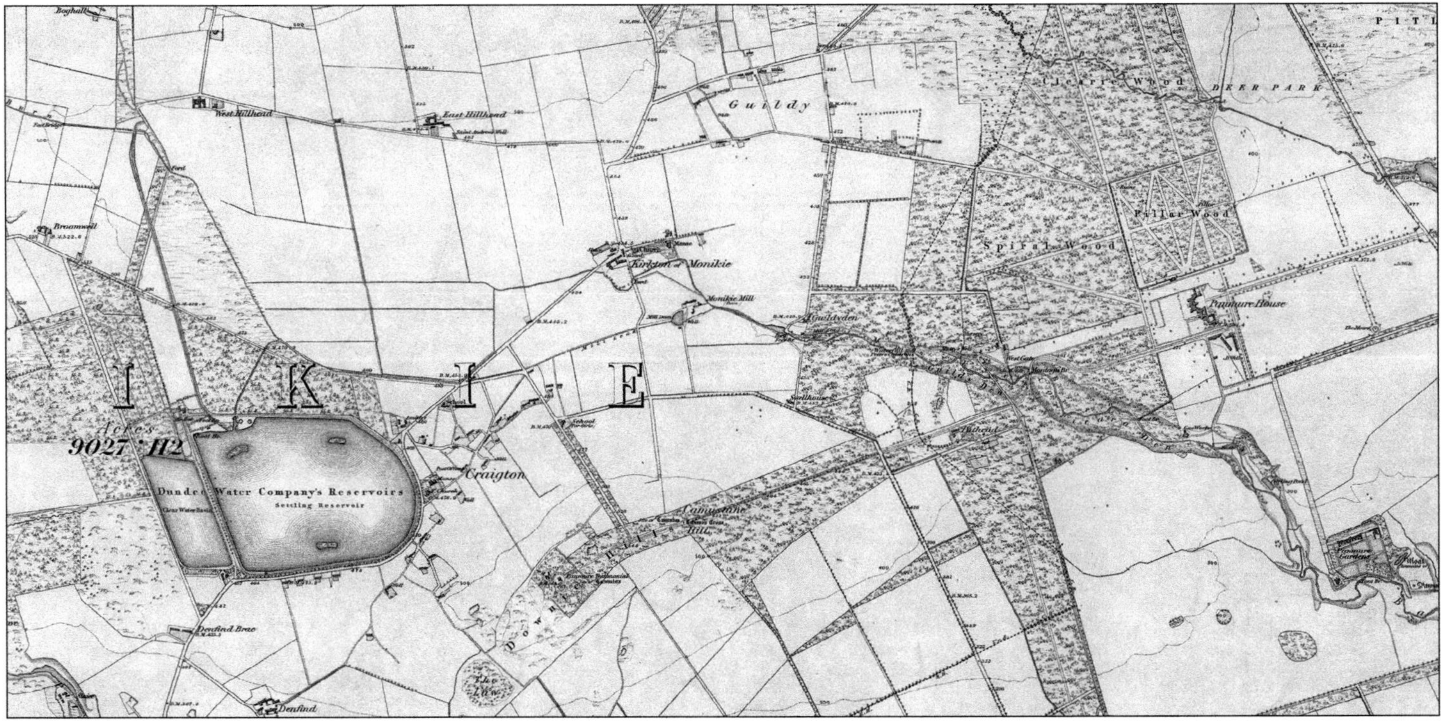

Plate 5 Monikie in the 1850s. Detail of Ordnance Survey, 1st edition, 6 inches to the mile, Forfarshire Sheet L1, surveyed 1858.

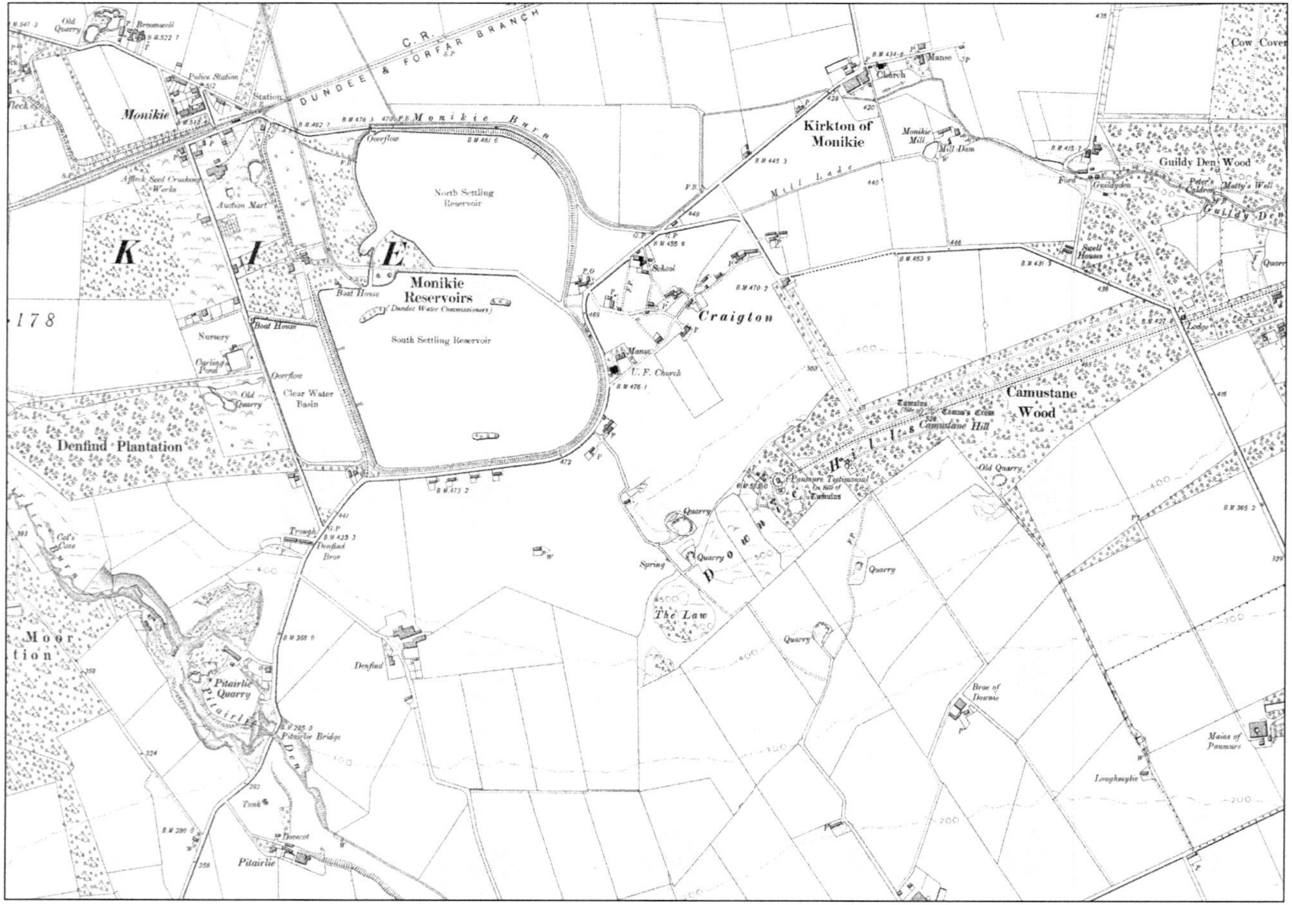

Plate 6
Monikie in the 1900s. Detail of Ordnance Survey, 2nd edition, 6 inches to the mile, Forfarshire Sheet LI.NW, surveyed 1900.

From at least the beginning of the sixteenth century extensive gardens were being laid out with walled enclosures, statues and garden rooms modelled on European lines, like the Great Garden at Edzell Castle in Angus. In the eighteenth century aristocratic owners began to adopt English-style tree-lined avenues, waterworks, parterres and walled kitchen gardens.[33] The minister of Duddingston, near Edinburgh, described the 'picturesque, elegant ... attracting' and highly fashionable features of the earl of Abercorn's grounds in typical and typically flattering (Church of Scotland clergy were appointed by the local landowners) terms in the 1790s. He listed the shrubbery, pleasure grounds, canal, lakes, isles and cascade, the clumps of trees replacing the old-fashioned formal avenues, and admired the 'grandeur and beauty of the ... prospects'.[34] Landscape gardening never caught on in Scotland to the extent it did in England, but, just as south of the border, the fashion for gardening took hold of the gentry and the rising urban middle class who bought country estates:

> The house of moderate size, unpretentious in architecture, with its walled garden and nearby orchards, with its lawn shaded by fine trees, with surrounding parks long laid out in grass, with its belts of timber sheltering the polices, the whole domains often surrounded by a stone wall ... is typical of a Scottish country house.[35]

The nineteenth century saw the gardens of middle- and upper-class Scots follow the fashions and taste of England as landscaping gave way to Italianate, neo-Tudor and then wild or woodland gardens. But through all these passing trends, a central purpose of the garden was to reflect its owner's status and aesthetic sense. This was 'a world in which wealth, taste and acceptability was measured out in flower gardens and orchid houses, asparagus and grapes'.[36]

Not so the gardens of rural workers. These did not often attract the eye of passers-by and, as we shall see, judgements were at best mixed. Cottage gardens did, however, just like those of workers' masters', reveal the character and social position of the inhabitants.

THE MORAL LANDSCAPE

The landowners who brought into being the man-made landscapes did so primarily for economic reasons, but they had other concerns too. Beauty in the arrangement of their gardens, their policies and the vista across the countryside from the mansion was important not only for itself but for the prestige and

33 McKean, *The Scottish Chateau*; Brown, *Scotland's Lost Gardens*.

34 *OSA* Duddingston, Midlothian, vol. XVIII, 364.

35 Cox, *Gardening in Scotland*, 96.

36 Morgan and Richards, *Paradise out of a Common Field*, 11.

power it bestowed, or implied. Cottages and their gardens played their part in the view from the house and their location and appearance had therefore to be carefully considered. David Gorrie, whose many articles in the mid-nineteenth century urged the case for ornamental as well as better-built workers' cottages, wrote that cottages 'may enliven a scene otherwise dull – a scene having no inviting distance, and no striking feature in the foreground'.[37] Cottage design, he believed, should not attempt to mimic that of the mansion or be fanciful or extravagant but should harmonise with its style to create a unified perspective.

Gorrie had less to say about the inhabitants of these picturesque dwellings, who would of course have been highly visible in the landscape their labours had created. The prevailing upper-class view of the rural working class in Britain and Europe, through its several transformations, was represented in the changing preoccupations of the landscape art, and literature, of the time. French artists such as Claude Lorrain, Poussin and Watteau depicted rural life in the seventeenth and early eighteenth century as a classically inflected pastoral idyll peopled with elegantly entwined shepherds and shepherdesses. Dutch painters such as David Teniers the younger and Jan Steen showed the cheerier, bucolic side of rustic society.

In England, as John Barrell points out in *The Dark Side of the Landscape*, it was the poets not the painters who took up the pastoral tradition, in which 'no moral value is ascribed to industry', and love came before work. At the end of the eighteenth century, however, this image was chased away by the French revolution: quite suddenly the poor became the mob – terrifying, unpredictable, hostile. Gainsborough's paintings and Crabbe's poems depict rural workers controlled by their work: labouring in a productive and efficiently managed landscape, engaged in rural trades or returning exhausted to their families. Drinking together in the village pub, dancing with the young squire or relaxing on the green became subversive.[38] As the popular painting *Village Politicians* (attributed to David Wilkie) shows, in such settings the people were likely to be indulging in ignorant and therefore potentially dangerous political talk.

Constable, today still one of the most popular British artists, took what Barrell calls the repression of the poor a stage further by hiding the rural workforce in the distance, where individuality, expression or attitude can no longer be seen and where 'the labourers do not step between us and the landscape – they keep their place, and it is a very small place, a long way away'.[39]

The 1850s and 1860s saw a shift in British middle-class attitudes to the poor, including the rural poor, and this is reflected in the way they were now portrayed in art. Scottish landscape painters, or those not in thrall to the Highlands,

---

37 D. Gorrie, 'Cottages in park scenery', *NBA*, 24th October 1849, 10.

38 John Barrell, *The Dark Side of the Landscape: the rural poor in English painting, 1730–1840*, Cambridge 1980, 10–11, 66–83.

39 Ibid., 104.

were depicting ploughmen in bleak fields, harvesting, women lifting potatoes, tumbledown cottages, muddy tracks and patches of wasteland, following the example of Millet and Corot in France. Cottages were often shown with an old woman outside the rose-entwined door, suggesting tranquillity but also inevitable rural decline. John Morrison has pointed out that, 'as urbanization and industrialization continued to develop very rapidly, rural images in Scotland between 1845 and 1865 often carried nostalgic, elegiac messages.'[40] Later, however, they were more likely to foreground the worker and challenge the viewer. In the naturalistic and unsentimental paintings of Arthur Melville and the Glasgow Boys, strongly influenced by Van Gogh, Millet and Bastien-Lepage, workers appear in the foreground of paintings of rural scenes as individuals, not concealed in the landscape and not beaten down by their labour.

Postcards too, now hugely popular, provided views of (rather clean) village streets, rural scenes and lonely cottages, often peopled with women and children in their best clothes. Valentines of Dundee and George Washington Wilson in Aberdeen specialised in this type of scene – almost as popular as fisher folk – apparently photographed for no other reason than their rustic quaintness. Front walls completely covered in creepers, tiny flower beds against the walls and gardens filled with flowers are frequent, attesting to the continuing taste for the English cottage garden amongst the postcard-buying public.

With the exception of the mining villages, which were to a great extent hidden from view and under the eye only of the colliery owner, every aspect of the rural worker's life was exposed to the gaze of his betters. The cottage often stood beside the public road, where it could be inspected by every passer-by. Unlike workers' quarters in towns and cities, even dwellings tucked away down farm tracks or half-hidden on the edge of woods and muirs were accessible to the great and the good of the parish, from the minister to the soup-and-pamphlet-bearing countess. And, although it was recognised that it was the landowner who had the chief responsibility for providing a comfortable dwelling, it was the responsibility of the worker's family to ensure that their home and their garden did not disturb the harmony or the pleasing fecundity of the rural scene, whatever the cost in time and effort. Neat gardens in front of well-kept cottages civilised and so enhanced the whole landscape.

[40] John Morrison, *Painting Labour in Scotland and Europe, 1850–1900*, Abingdon-on-Thames 2014, 6.

CHAPTER 2

KAILYARDS AND FARM SERVANTS

'At present there is too much of the "kale yard" about a hind's [farm servant's] garden', wrote R. Hunter Pringle, concluding, as Assistant Commissioner, his report on agricultural workers in the Lothians for the 1893 Royal Commission on Labour.[1] What he meant was a very limited range of crops and a clinging to traditional modes of cultivation. This chapter describes these gardens, the direct descendants of the ancient Scottish kailyards, provided, along with the house, as part of the farm servant's wage.

The criticism implied in Pringle's choice of words is deliberate. By the end of the nineteenth century the term 'kale yard' or 'kailyard' (the older spelling) had become indisputably negative, not least because of the 'Kailyard School', a disparaging label applied to the hugely popular works of J. M. Barrie, S. R. Crockett, Ian Maclaren and others for their sentimental depictions of rural life stuck in an unchanging and parochial past.[2]

The words for garden in all European and Slavic languages derive from roots that mean 'enclosure'. 'Yard' itself is related to Indo-European 'gher' or fence, and thence to Old English 'geard'.[3] Kailyards in Scotland go back to the Middle Ages and perhaps further,[4] and can still be found today in Shetland (Plate 7). Anyone, rich or poor, landowner or cottar, could have a kailyard for raising vegetables before the era of improvement. They were defined by George Skene Keith in 1811 as 'a small enclosure of a few falls of ground for planting coleworts [kale] or raising

1 *RC Labour*, vol. III, pt II, Reports by R. Hunter Pringle and Edward Wilkinson, London 1893, Edinburgh and Haddington, 122.

2 Andrew Nash, 'The kailyard: problem or illusion?' in Susan Manning, Ian Brown, Thomas Owen Clancy and Murray Pittock (eds), *The Edinburgh History of Scottish Literature*, Edinburgh 2007; Roderick Watson, *The Literature of Scotland: the Middle Ages to the 19th century*, London 2007.

3 John Dixon Hunt, 'Introduction', in J. Dixon Hunt and J. Wolske-Bulmahn (eds), *The Vernacular Garden – Dumbarton Oaks Colloquium Series in the History of Landscape Architecture*, Washington 1993, 19.

4 It is possible that kale was known in the far north of Scotland and the Northern Isles in Neolithic times (Ian Tait, personal communication, 29th June 2016).

Plate 7 Shetland kailyard. Author's photograph, 2014.

garden roots'.[5] The enclosure – a dyke, hedge or fence – was essential to protect the intensively cultivated kailyards from free-ranging cattle and children. On the unenclosed infield and outfield land farmers relied on small boys – the farm servants' sons – to keep livestock and other predators off the growing crops.

Kailyards were important. Barony (local) courts in the seventeenth century set regulations to ensure that tenants and sometimes cottars had kailyards for greens and that they protected them with dykes.[6] Tenants were also frequently obliged to plant ash trees around the kailyards.[7] The range of crops was always limited. Farmers were even discouraged from attempting to grow vegetables other than the most basic. Lord Belhaven's *The Countrey-man's Rudiments, or, An advice to the farmers in East-Lothian, how to labour and improve their ground*, published in 1699, counselled his readers to raise 'nothing but what is both profitable and easie': potatoes and turnips and, 'all that is necessar for Herbage' – viz. cabbage and summer kale. 'Leeks, Sybous [spring onions] and other curious Herbs not being for your handling'. These vegetables were to be grown in half an acre of ground on the north side of the farmhouse, 'where you may have a little door to go in at, by which it will be preserved from the Summer scorching Sun, and Southern Wind, the two greatest enemies to Kitching Grounds'.[8]

5 George Skene Keith, *GV Aberdeenshire*, Edinburgh 1811, 138.
6 Ian Whyte, *Agriculture and Society in 17th century Scotland*, Edinburgh 1979, 59.
7 *OSA* Wemyss, Fife, vol. XVII, 525; *NSA* Birnie, Morayshire, vol. XIII, 85.
8 Lord Belhaven, *The Countrey-man's Rudiments, or, An advice to the farmers in East-Lothian, how to labour and improve their ground*, Edinburgh 1699, 32.

A short story by Alexander Bethune describes an old fermtoun and its kailyards:

> There they [the cottages] stood with their low walls, built in some instances with clay instead of mortar – roofs composed of alternate layers of thatch and turf – chimney tops with a rope of twisted straw around them to keep them together, and doors so low that their inhabitants were obliged to 'look low' before they could enter. Another feature of days departed was the little gardens in which were cultivated small quantities of cabbages and potatoes, surrounded by what, to a stranger's eye, or indeed to any eye, might seem a nettle bank instead of a wall. The houses were disposed in no regular order, but stood in groups of two or three together, generally in the lowest places of an undulating surface, while the intermediate and higher ground was occupied as gardens in the way already noticed. Narrow green lanes formed the only communion between them, – sometimes crooked, sometimes straight, as the fancy of our forefathers had been.[9]

As interest in the garden grew amongst the gentry in the eighteenth century and began to attract the middle classes, including the rapidly prospering farmers, the kailyard became more and more the preserve of the cottar and then the farm servant. Stocked with potatoes and kale and the other ingredients of broth – cabbage, turnips (swedes – 'neeps' in Scots), sometimes with onions and leeks – and little else, this plot – functional, utilitarian, lacking ornamentation, trees or much in the way of perennial planting – lasted until well into the twentieth century. A miner's daughter brought up in Fife in the 1930s remembered that there were no flowers – 'you can't eat flowers' – but early potatoes, lettuce, leeks and kale. Father delved and mother weeded.[10] As recently as the 1970s a farm cottage-garden in Fife, in front of the house, retained all the traditional features of a kailyard: planted only with vegetables, no flowers, no fruit bushes and no trees.[11]

In associating the kailyard, or kailyard style, with farm servants, it is not intended to suggest either that all farm servants kept this kind of garden or that all kailyards belonged to farm servants. But, for reasons that this chapter will explain, kailyards became in the nineteenth century the typical garden of the farm servants with little free time, not much cash, living far from villages with shops and markets, and no security in their job or their home.

---

9 Alexander Bethune, 'The Fate of the Fairest', in Alexander and John Bethune, *Tales of the Scottish Peasantry* [1843], London 1884, 61.

10 Personal communication from Mrs Margaret Wallace, 12th October 2015. Her father, 'a great one for trying to improve our lot', also bred guinea pigs in huts in the garden to sell to medical researchers.

11 Personal communication from Helen Hammond, 7th July 2017.

FROM COTTARS TO FARM SERVANTS

Early eighteenth-century estate plans show that, amongst the higgledy-piggledy layouts of the fermtouns, most of the tenants, subtenants and cottars had kailyards, sometimes shared amongst two or three houses. Like the Shetland yards, they were attached to the tenants' houses, where they could most easily be dunged from the byre and the midden and where the dykes could be easily maintained.

George Robertson, farmer and writer on agriculture, recalled the old Midlothian farmhouses of his youth in the 1760s at the start of the epoch of improvement, with the 'sit-house' or farmer's dwelling house in the middle, barns at one end and the byre at the other. Behind this 'farmer's ha'' was the kailyard, 'which was cropped with cabbages, a few, and with open kail, a great many'.[12] Cottars, subsisting on day labouring and the produce of a few rigs, lived in small low buildings made of turf and rounded field stones, roughly thatched. Their kailyards kept them in vegetables all year round, so that the daily broth (before potatoes) never lacked the leafy vegetables such as cabbage, kale and leeks known as potherbs.[13]

In the old fermtouns nearly everyone had some land, whether tenants, subtenants or cottars. Subtenants and cottars also worked as day labourers or as artisans – weavers, smiths, wrights, shoemakers, tailors and so on. Although the two groups were often not differentiated in contemporary sources, subtenants had tacks (written or verbal leases) from tenants and rights of tenure under Scots law, while cottars were allowed small patches of land, often less than an acre, in return for their labour, did not pay rent and could be evicted without formal proceedings.[14] T. M. Devine has characterised them as 'among the most shadowy groups in seventeenth and eighteenth century Scottish social history'.[15] Cottars probably made up around one-third of the rural population in 1700,[16] but by the nineteenth century they had disappeared as a class: some emigrated to nearby towns and mills, the more adventurous to North America, others became casual labourers. Many became full-time waged farm servants – 'hinds' in the south-east. Confusingly, these last were often referred to as 'cottars' well into the nineteenth century, but in fact farm servants were landless, totally dependent on their wage and the farmer, the 'master'.

The farm servants' gardens need to be understood in the context of the working and living conditions of these ploughmen, cattlemen, shepherds and orramen and their families who tended them. The terms of the servants' hire,

12 George Robertson, *Rural Recollections; or, the progress of improvement in agriculture and rural affairs*, Irvine 1829, 78.

13 Ibid., 69.

14 Devine, *Transformation of Rural Scotland*, 136.

15 Devine, 'Social Responses to Agrarian "Improvement"', 155.

16 Ibid., 140.

their wages, their hours of work, their families' involvement in the work of the farm, their relationship with the farmers and the tied cottage all had an impact on the garden and on how they managed it.

The farm was a small and largely self-contained world, and became more so through the century. Its social and emotional life, dominated by the characters of the farmer and his wife, the attitude of the grieve and the bustle of the farmyard made for a powerful brew, especially for the bothy-bound unmarried servants, living on the farm. The married servants and their families had some escape from this after working hours, but must often have felt imprisoned in their carefully distanced cottage rows.

In the mid-eighteenth century, even while new techniques and farm amalgamations were already in train in the Lothians, the distinctions between farmers and servants were slight. The gudeman's family and his unmarried servants still lived cheek by jowl and ate the same food together at the same table. The house of the gudeman was bigger but not much more comfortable or better furnished than those of the cottars; both wore homespun and woven cloth; both read the same mainly religious books.[17] With greater profits from farming came greater social distance between master and servant. The *General Views*, written during the French and Napoleonic, when wages had risen sharply in most places – they had doubled in fifteen years in several counties – are generally appreciative of the intelligence, industriousness and respectability of servants and labourers. Robert Somerville, a surgeon who wrote the *GV* for East Lothian, seeking to explain the 'regular and perfect manner in which all rural operations are performed in this county', cites the expertise of the labourers and the 'well-regulated' relationship between master and servant.[18]

But in the 1850s, when wages were once more on the rise and Perthshire ploughmen began an agitation for a ten-hour day, letters to the press urged them to concentrate on higher wages, getting rid of the bothy system, their own education – anything but reducing hours of work. The appeal to an idyllic past was thrown in: 'Let them be to a greater extent incorporated with the master's family, and so treated as to feel that the austerity of the master is sunk in the kindliness of the father and the friend … their morals [will] be safe … and their spiritual welfare attended to', wrote one farmer to the *Perthshire Advertiser*.[19]

The general view of the Scottish farm servant in the later nineteenth century was one of patriotic admiration, especially in the south-east. Robert Scot Skirving, an East Lothian farmer and a leading figure in national agricultural circles, wrote in a widely distributed pamphlet of the great superiority of East Lothian ploughmen over their 'lifeless, soulless' English counterparts, overpaid and underworked, 'who do not take it as an insult to be offered prizes for

[17] Smout, *Scottish People*, 305–7.

[18] Robert Somerville, *GV East Lothian*, Edinburgh 1813, 206.

[19] *Perthshire Advertiser*, 13th December 1855, 3.

moral character at agricultural shows'.[20] Scots farm workers were more productive than the English – to the tune of £20 per year per head, even though the soil was poorer. They were also more respectful, more sober and better educated – and much credit was due to payment in kind, which discouraged 'dissipation'.[21] In the 1890s, Angus and Perthshire workers were 'sturdy, sensible, and on the whole contented'[22] and 'sober, generous and industrious'; in Aberdeenshire they were intelligent and 'physically excellent'; and in Fife and Lothian men were 'honest and truthful'.[23] These testimonials came at a time when farm workforces could be over a hundred strong, and one wonders how farmers were able to characterise so many men and women so authoritatively.

One can detect more than a hint of nostalgia in Assistant Commissioner Pringle's urging the farmers to plant fruit trees and bushes in labourers' gardens: this would not only improve the lives of the workers but was also a cheap way to promote better relations between masters and men. The withering away of the paternalism of the masters and the compliance of the men was a frequent lament.[24]

The farm servants of Perthshire and Aberdeenshire in the later decades of the nineteenth century could tell a different story. Frequent flitting, the rise of trade unionism in this region and not least the insults hurled at farmers in the bothy ballads all bear witness to poor relations between masters and men. The social gap between them had widened to a chasm. Ian Carter points to the practice of farmers in Aberdeenshire buying food for their own families from the shops in Aberdeen, while their servants were fed on the oats, potatoes and milk produced on the farm, as symbolising the width of the social gap between the two.[25]

They'll tip you on the shoulder
    And speir gin ye're to fee; [ask if you're looking for a hire]
They'll tell ye a fine story,
    That's every word a lee;
They'll tell ye a fine story,
    An' get ye to perform;
But lads, when ye are under them
    Ye'll stand the raging storm.

The Weary Farmers[26]

---

20 Skirving, *Landlords and Labourers*, 98.

21 *Journal of Agriculture*, 1851, 447.

22 *RC Labour*, vol. III, pt I, Reports by H. Rutherfurd and G. R. Gillespie, 1893, Forfarshire and East Perthshire, 147.

23 *RC Labour* vol. III, pt II, Summary Report, 36.

24 Ibid., 61.

25 Carter, *Farm Life*, 23.

26 John Ord, *Ord's Bothy Songs and Ballads of Aberdeen, Banff and Moray, Angus and the Mearns* [1930], Edinburgh 1995, 212.

As a 1906 government report put it, 'the old familiar semi-patriarchal relationship under which the labourer was partly bullied and partly petted, was replaced by a more commercial spirit, the tie between master and man became almost exclusively a "cash nexus"'.[27]

THE LONG HIRE

The account of the farm servant's life and work that follows is of necessity a summary of a highly complex system that varied from farm to farm as well as between counties and regions, and over time. The main features of the farm service system were to be found throughout the Eastern Lowlands, however, and across the border into Northumberland.

The 'long hire' – a year's contract for married servants, six months or a year for the unmarried – was a key element of farm service. The new breed of large tenant farmers who emerged during the agrarian revolution held leases that bound them to farm the land in specific ways. Long rotations (five or seven years) included new crops such as turnips and sown grasses and these extended the farming year, demanding a large year-round permanent workforce as well as seasonal labourers and their families. Turnips, for instance, had to be hoed for six months of the year. Horses were now used for the skilled work of drilling as well as for ploughing and carting. It was this changed pattern, together with larger cattle herds, that required the hiring of permanent full-time workers.

Married servants needed cottages on or near the steading. The long hire and the tied cottage for married servants with a piece of land were not new – they dated from at least the sixteenth century, notably in the south-east.[28] Unmarried, mostly younger, servants lived in the farmhouse kitchen and later in the notorious and comfortless bothies (where the opportunities they provided for immorality and the moral neglect that they represented absorbed the energies of ministers and reformers for decades).

The long hire increased the control the farmer had over his men and their families: the servant could not leave his place until the end of his term.[29] But at the term the hire could end without warning, on either side. With the loss of the job went the man's home and his garden.

At the top of the farm hierarchy was the grieve or foreman. Under him were the ploughmen or horsemen, the largest and most specialised group – the apprenticeship was two years – who usually commanded the highest wages

27 Board of Agriculture and Fisheries, *Report on the Decline in the Agricultural Population of Great Britain 1881–1906*, London 1906, 10.

28 Devine, 'Scottish farm service', 1.

29 At Whitsun (May) for a one-year hire or Martinmas (November) in the case of a six-month hire.

Plate 8 Grieve and ploughmen with horses at Craigie farm, Leuchars, Fife, c. 1890.

(Plate 8). Cattlemen came next, with the orramen, the general workers, at the bottom. Shepherds, who lived and worked far from the steading, had a degree of independence and their own flocks. Farm workforces expanded: towards the end of the century large farms in the south-east were employing over a hundred servants and labourers, men and women, with more during harvests.[30] Working long, arduous hours in all weathers, horsemen needed to be young and fit: eighty per cent of Aberdeenshire horsemen in the 1901 census were aged between fifteen and thirty-five.[31] As the ploughmen aged and their strength started to fail, they were most likely to become orramen or labourers. Only a few would be promoted to grieve. Cattlemen and shepherds, whose work was less arduous, lasted longer as hired servants. With little chance of promotion or savings to take on a croft or smallholding, except in the north-east, most would have to leave their farm cottage and garden when disabling illness or old age struck, although married servants in the south-east at least did have sick pay – by law.[32] Most moved off the land and found labouring jobs in towns and villages or might have the chance to set up as carters or run dairies; some emigrated to North America.

30 Alastair Orr, 'Farm servants and farm labour in the Forth valley and South-east Lowlands', in T. M. Devine (ed.), *Farm Servants and Labour in Lowland Scotland 1770–1914*, Edinburgh 1984, 39.

31 Carter, *Farm Life*, 78.

32 Orr, 'Farm servants', 36.

The farmer's practice of sacking men at the term, often for no reason, or, on the other side, the servant's of upping sticks and moving to another farm, also often for no reason, became ingrained. More is heard, inevitably, in contemporary accounts, of the servants' irritating 'frequent flitting' than of the farmer's casual high-handedness. The prosperous cereal district of Berwickshire was supposedly most cursed by this, but it was well known throughout the Lowlands. Contemporary observers were perplexed, as gains in wages or conditions were often marginal or non-existent. There were sometimes practical reasons – to gain experience, to find a better employer or better accommodation; or social – to move closer to the school and the church or to relatives. Flitting for some did become a habit, there is no doubt.[33] One frequently expressed complaint was the tendency of farm servants' wives to fall out with their neighbours, perhaps not surprisingly given the isolated communities they lived in.

Whether the autocratic farmer or the itchy-footed ploughman was to blame, the sight of carts laden with children, pots and pans, perhaps a dresser and, in earlier times, a door and window frame trundling along the lanes from one farm to another at the term in late May was a familiar one. On the other hand, many servants stayed with the same master for years, but there was no security, no guarantee that the master would not suddenly take against a loyal servant.

The result for the garden was inevitable: investment of labour or effort in the soil was a waste of time and effort – what the family needed was as high a crop yield as possible. And, when the servant left, he received no compensation for any crops left standing.

PAYMENT IN KIND

It is the Scottish farm servants' wage system that explains the role of the gardens most starkly. In the eighteenth century the bulk of the married servant's wage was paid in kind. This generally was made up of the house and garden, six and a half bolls of oatmeal (one boll of oatmeal weighed 140lb), a cow's keep (grazing) or three pints of milk a day, fuel (coal or peat), and manure for the garden. The money wage might be less than £6 a year and in East Lothian some were still being paid entirely in kind in the early years of the nineteenth century: as well as oatmeal (ten bolls) they had barley, dried peas, flax seed and meals during harvest and had to pay only for shoes, although these too were sometimes included in the wage. They paid for their children's education by selling their surplus meal.[34]

In the south-east flax seed was often part of the wage, but it might be grown in a corner of the farmer's fields rather than on a separate piece of land. This

33 Rab Houston, '"Frequent flitting": geographical mobility and social structure in mid-19th century Greenlaw', *Scottish Studies*, vol. 27, 1983, 27–42, 38.

34 Somerville, *GV East Lothian*, 206–8.

practice was dying out in the early 1800s and the land granted by the farmer used for potatoes.[35] Potato grounds, as they were called, became customary and highly valued (see Chapter 4).

Bad harvests in the 1790s and the end of the Napoleonic Wars in 1815 saw money wages plummet as grain prices fell.[36] But payment in kind cushioned servants from their worst effects: 'a hind here ... has no care upon his mind, no rent to pay, no bad seasons to dread, for whatever may be his master's crop, he is sure of his full share ... ', wrote an East Lothian farmer, perhaps rather bitterly.[37] The garden crops might suffer in a bad harvest year, of course, but it was recognised that farm servants were better off than artisans and labourers, who had to subsist entirely on their wages and the gardens for which they often had to pay rent. By the 1840s married servants outwith the north-east were still receiving most of their wage in kind, but this varied from eighty to ninety-five per cent of the total wage in the south-east to just over half in the central counties of Fife, Angus and Perthshire.[38]

The period of high farming after 1850 did not immediately lead to a better standard of living for farm workers. It was not until the agricultural depression of the 1870s that wages started to rise and food prices to fall. By the 1890s ploughmen's wages amounted to £45 in the north-east and £50 further south;[39] by this time, agricultural wages in Scotland were above the British average.[40] Payment in kind had been abandoned completely on a few farms but East Lothian hinds did not get the bulk of their wages in cash until the 1880s, and a quarter of them were still being paid partly in kind in the twentieth century.[41] Further north, the same amount of oatmeal and milk was still provided as before, although milk could come in 'very very small pints'[42] and common complaints related to the quality of both the milk and the meal. Milk and potatoes as part of farm workers' pay survived until the 1970s in Angus.[43] Tied cottages, albeit with increasingly strengthened rights for workers, also survived. The garden too, therefore, remained part of the wage.

---

35 Sir John Sinclair, *An Account of the System of Husbandry Adopted in the More Improved Districts of Scotland, with Some Observations on the Improvements of which they are Susceptible*, Edinburgh 1812, vol. II, 132–3.

36 Orr, 'Farm servants', 32.

37 Somerville, *GV East Lothian*, 209.

38 Ian Levitt and T. C. Smout, 'Farm workers' incomes in 1843', in T. M. Devine (ed.), *Farm Servants and Labour in Lowland Scotland 1770–1914*, Edinburgh 1984, 172, Table 2.

39 *RC Labour* vol. III pt I, Aberdeenshire and Kincardineshire, 130; Ibid., pt II, 105.

40 Jeanette M. Brock, 'The importance of emigration in Scottish regional population movement 1861–1911', in T. M. Devine (edn), *Scottish Emigration and Scottish Society*, Edinburgh 1992, 115.

41 Orr, 'Farm servants', 36.

42 Robert Dempster, *The Natural Horseman, the horse and ploughman*, Dundee 1884, 10.

43 Personal communication from Aileen Elliott, 5th December 2019.

THE WORKING DAY

Long hours of work were recognised by some contemporary observers as a real deterrent to gardening. Work for men and women took place outside in all weathers and seasons and in the barn in the dark mornings of winter. The cattleman's hours were the longest but horses needed grooming and 'suppering', all done by the ploughman at the end of the day and until well after nightfall. Sunday was the only day of rest until the very end of the nineteenth century, and horses and cattle still needed to be fed on that day.

Hours seem to have fallen very slowly but fairly uniformly across the Lowlands. Alexander Wight's account of a farm he visited in Strathmore, Angus, in the 1780s tells us that the servants started work at 4am, took one hour off for breakfast and another for dinner, and did not finish until 8pm – a sixteen-hour day. In the Lothians the working day was from 5am to 7pm, with two hours' rest.[44] Hours in Aberdeenshire were similar. By the 1790s hours in the Borders were 'more gentle' – 6am to 6pm in summer, including two hours' break but with an additional two or three hours' suppering[45] – agricultural writers often ignored this part of the working day. Winter hours were shorter, but there was always the 'barn work' of threshing and winnowing grain for the mill to keep the men busy. The six or eight weeks of harvest meant working hours were even longer. Thus the time of year when most work was needed in the garden, spring and summer, was also the time when the hours of work were reaching their peak.

By mid-century not much had changed. In Aberdeenshire fifteen hours a day was the norm, from 4am to 7pm in summer.[46] It was not until almost the end of the century that ten hours (plus, each day, over three hours for grooming and suppering) became usual. But William Diack, a former farm servant, writing in 1905, maintained that between March and October a ploughman still worked from five in the morning until nine at night.[47] Farmers resisted the reduction in hours demanded by the ploughmen and their unions in many places. On this point the Royal Commission on Labour Assistant Commissioner for the northern counties wrote: 'On the whole there is very little complaint of want of holidays [six days and the odd cattle show per year] or of long hours. That the hours are too long for health is, to anyone who has seen the race of farm servants, a manifest absurdity'.[48] Clearly, only health mattered.

Could work be done in the garden on Sundays? The Sabbath was strictly observed in earlier times and kirk sessions, the committees of elders who regulated the parish, could impose heavy penalties for breaches. There was

---

44 Andrew Wight, *The Present State of Husbandry in Scotland*, Edinburgh 1784, vol. IV, 27.

45 Alexander Lowe, *GV Berwickshire*, Edinburgh 1794, 54.

46 Carter, *Farm Life*, 111.

47 William Diack, 'The Scottish farm labourer', *The Independent Review*, vol. VII, 1905, 315–27, 317.

48 *RC Labour*, vol. III, pt I, Moray, Banff and Nairn, 111.

the notorious case of Robert Burns' friend Gavin Hamilton, who was fined by Mauchline kirk session in Ayrshire for ordering new potatoes to be dug up from his garden on a Sunday (part of a concerted campaign against Hamilton, who eventually won an acquittal).[49] But otherwise it has not been possible to find evidence of kirk sessions, whose powers to police behaviour were waning in the nineteenth century anyway, attempting to prevent or even discourage Sunday gardening – or of clerical or other expressions of disapproval of the practice. Evidence may remain to be uncovered, of course, but had this been happening generally, farm servants' gardens would have been in an even worse state.

The man's work and his hours of labour tell only part of the story. Wives and children were expected to help out at busy times of year and were part of the contract between the servant and his master in the Lothians and the Border counties. Here the rent of the house was covered by the labour of the 'bondager', a woman from the household, a wife or a sister or daughter, at busy times – seed-time, turnip hoeing and the harvest.[50] Until the 1872 Education (Scotland) Act, many children were working in the summer from as young as seven: a young boy would be set to bird-scaring or cattle-herding before he was taken on as a halfling ploughman or orraman on the farm.

FARM COTTAGES

During most of the nineteenth century the majority of farm cottages were badly built, poorly maintained, damp and overcrowded (Plate 9). One-room earthen-floored dwellings were still to be found at the end of the century. The old cottar houses had mostly been built of turf and clay or clay and bools (rounded field stones) and thatched with turf or heather. They stayed warm and dry only if constantly maintained – otherwise they soon became leaky and miserable. In the north even the newer houses in the early 1800s were being built of stone and clay and thatched with straw. The floors were beaten earth, there were, at most, two very small glazed windows and no ceilings.[51] In the more advanced farming counties further south new cottages had walls of dressed stone, the occasional ceiling and wooden floors, but they were still small (sixteen by eighteen feet).[52]

The postwar economic downturn and rising unemployment meant that building and repairs slowed or ceased after 1815. Cramped, cheaply built and poorly maintained cottages soon degenerated into rural slums. It was not until

49 Future Museum.co.uk, South-West Scotland, http://www.futuremuseum.co.uk/collections/people/key-people/burns/gavin-hamilton.aspx, accessed 22nd November 2020.

50 M. Robson, 'The Border farm worker', in T. M. Devine (ed), *Farm Servants and Labour in Lowland Scotland 1770–1914*, Edinburgh 1984, 82–3.

51 Skene Keith, *GV Aberdeenshire*, 138.

52 Robertson, *GV Midlothian*, Edinburgh 1795, 42.

Plate 9 A hind's cottage at Norham, Northumberland. W. S. Gilly, *The Peasantry of the Border: an appeal on their behalf*, Berwick on Tweed 1842.

the panic engendered by the rural riots in England in the early 1830s that a few of the more progressive Scottish landowners, egged on by concerned clergymen and others, began to take an active interest in the condition of farm workers' housing, attempting to encourage or shame their fellows into building better cottages. But not all: a pamphlet published in the 1850s by a Carse of Gowrie ploughman condemned the damp and overcrowded cottages built 'where the least sacrifice of ground can be made'.[53]

In the 1860s and 1870s larger and better cottages started to be built, with privies and ashpits at the back, and many of these still stand today. Landowners and farmers had always pulled down cottages when their occupiers were no longer employed on the farm, and now fear of a rising parish poor rate (levied on property) or attracting 'paupers, pilferers and poachers' aggravated this.[54] The result was severe housing shortages in places, notably the north-east. As a result, many farm servants were forced to find accommodation in towns for their families while they slept in the bothy. Boys no longer grew up around the farm steading, becoming familiar with the routine and discipline of farm work.[55]

---

53 A Working Man, *The Clodpole: or the normal condition of agricultural labourers*, Dundee 1858, 6.

54 Skirving, *Landlords and Labourers*, 22.

55 James Alexander, 'Prize essay on the present condition of farm servants in Scotland with means to be adopted for its amelioration', *North of Scotland Gazette*, Aberdeen 1852, 6.

By the end of the century county councils were starting to take over responsibilities that had previously been the landowners', such as sanitation. The majority of workers, at least in central and south-eastern Scotland, were living in cottages built in the previous thirty years, drier and better constructed, if not much bigger, and, thanks to the work of County Sanitary Inspectors, equipped with privies. In fact, some recently built dwellings were, according to one observer, 'almost too good for the occupants',[56] although it was recognised that the best way to keep workers on the land was to provide good quality housing. Nevertheless, housing shortages remained, some families still lived in one-roomed cottages (for example, in Angus and Perthshire), while the floors in others were below ground level or too near the cattle court.[57] Twenty years later farm servants' accommodation was described by their union as 'a disgrace to the country'.[58]

How were such grim housing conditions relevant to the care of the garden? Families still had to eat, after all, regardless of the state of the home. A prevailing philanthropic view, shared by another ploughman, had the answer: 'let the labourer have a snug, comfortable cottage, however homely, with a piece of ground in a cheerful situation, and he soon becomes attached to the place while it exercises an invigorating influence over him in every respect.'[59]

THE VALUE OF THE COTTAGE GARDEN

The different elements of payment in kind held varying values for the worker – and for the farmer. Surplus oatmeal could be sold back to the miller, to neighbours or at a local market. This and the right to keep a cow or to three pints of sweet [whole] milk a day seem to have been the most prized items. The right to a cow pastured on the farm was jealously guarded in the south-east and in some other districts. Even more than the garden, it represented a vestige of independence, as any extra could be churned into butter or made into cheese and then sold. But increasingly milk was provided by the farmer as part of the wage and farm servants began to prefer this more labour-saving arrangement. By the end of the century it was mainly grieves who were allowed to keep a cow, herded with the master's cattle.[60]

For the farmer in the south-east at the beginning of the nineteenth century the grain allowance, much higher here than elsewhere, was reckoned to cost £22 per annum, the cow up to £10 and the land – garden, potato ground and flax plot –

---

[56] *RC Labour* vol. III, pt 1, Fife, Kinross and Clackmannan, 58.

[57] Ibid., 148.

[58] *The Scottish Farm Servant*, May 1913, 16.

[59] Alexander, 'Present condition of farm servants', 12.

[60] The value of the cow at this time was calculated as between £12 and £15 a year – around the same as a ploughman's money wage (*RC Labour*, vol. III pt II, 56).

no more than £3 10s.[61] The price of grain rose sharply during bad harvests – in 1740, 1783 and during the French and Napoleonic Wars and immediately postwar. It is conceivable that these events increased the value of the garden to both the hind and the farmer.

The cost to the farmer of the garden diminished in the latter part of the century. One Angus farmer told a government inquiry in 1893 that the cost of a farm servant's garden (manured) was £1 per annum, while the total cost of his wage was £52 6s. 1d.[62] The real value of the garden to the farmer was in attracting and keeping married hinds. Using data from the Scottish Poor Law Commission Report of 1843, Levitt and Smout have shown that only in the northern and north-eastern counties, at that time, was there a very clear preference for unmarried servants over unmarried. In lowland Angus and Perthshire, in around a third of parishes, married servants were preferred, while on either side of the Forth and in the south-east, in as many as two thirds of parishes, they held the advantage.[63] Levitt and Smout suggest that farmers in Angus and Perthshire in fact would have had married servants if they could get them.[64] What was the benefit, given the loss of land? Married servants were considered steadier and more dependable – 'more orderly and tractable'[65] – than the younger unmarried men, and brought the additional labour of their wives and children – cheap and sometimes free, in the case of the 'bondagers'.

The value of the garden to the worker is less easily established. Patrick Neill, the foremost authority on horticulture in the first half of the nineteenth century, urged farmers to provide larger gardens so as to attract married servants, which suggests that they were regarded as an asset. He claimed that some servants would prefer smaller money wages and good ground to more cash without land. How widely held this preference was we cannot know, but it does suggest that larger gardens were regarded as a considerable boon. And, where the wife and children can garden too, 'there can be a great increase in size'.[66] Nevertheless, although many, perhaps most, gardens were too small to be of much use and later in the century got even smaller, on the eve of the First World War a government report could still insist that the produce of farm cottage gardens 'is a considerable item in the family income'.[67]

---

61 Sinclair, *Account of the System of Husbandry*, vol. II, 136.

62 *RC Labour*, vol. III, pt 1, Forfarshire and East Perthshire, 156.

63 Levitt and Smout, 'Farm workers' incomes', 171, Table 1.

64 Ibid., 157.

65 J. F. Erskine, *GV Clackmannan*, Edinburgh 1795, 73.

66 Patrick Neill, *On Scottish Gardens and Orchards. Drawn up by desire of the Board of Agriculture*, Edinburgh 1813, 19–20.

67 *Royal Commission on the Housing of the Industrial Population of Scotland, Rural and Urban*, 1917–18, 170.

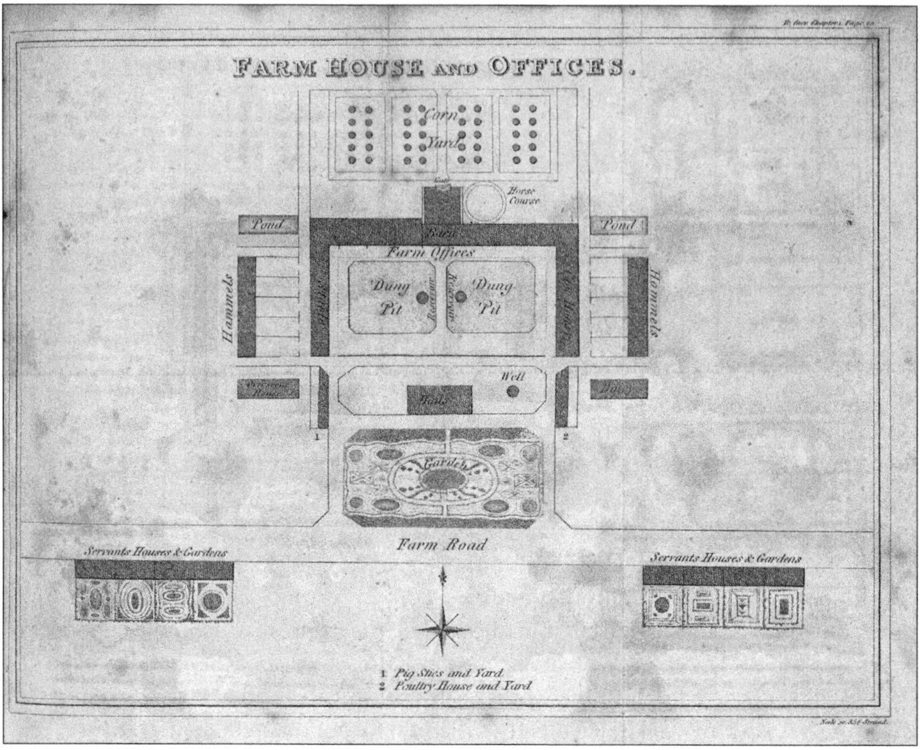

Plate 10 Plan of model farm, Sir John Sinclair, *Account of the System of Husbandry in Scotland*, Part 2, 20, Edinburgh 1812. The farm-servants' cottages and gardens are shown bottom left.

FARM COTTAGE GARDENS

In the old fermtouns houses were scattered about without any recognisable pattern, with the tenants and the cottars usually living quite close to each other, as we have seen, and close to the steading, with the kailyards generally right behind the houses. The increasing social distance between master and men came to be reflected in the physical distancing of the farm cottage from the farmhouse. As early as the 1760s, as George Robertson remembered, cottars' houses in Midlothian were usually clustered at a distance from the farmhouse.[68] Influential writers were soon recommending that cottages be 'at a convenient distance from the farm-stead, and so situated so as not to interfere with the farmer'.[69] Sir John Sinclair was more exacting: farm servants' cottages, except those of cattlemen, who needed to live close to the byre, should be built

[68] Robertson, *Rural Recollections*, 71.

[69] James Thomson, *GV Fife*, Edinburgh 1800, 380.

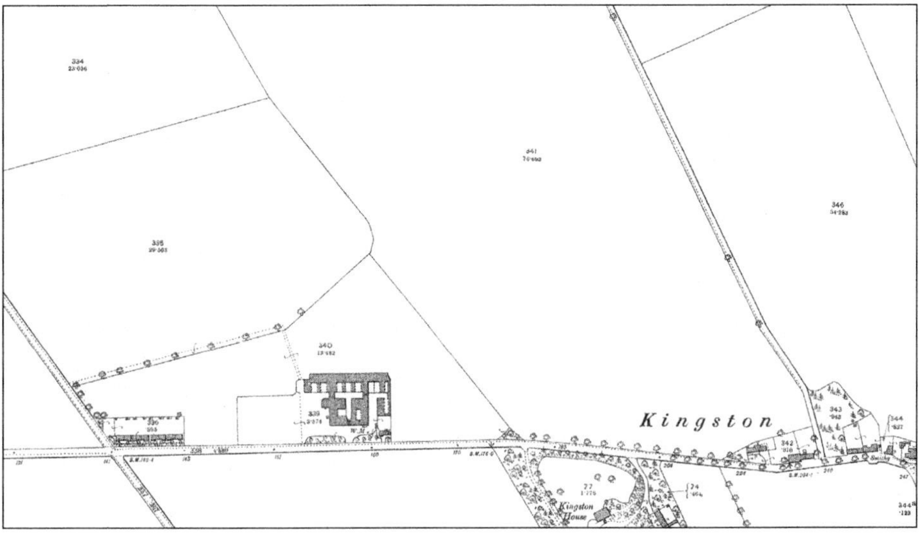

Plate 11 Kingston Farm 1893. Detail of Ordnance Survey, 2nd edition, 6 inches to the mile. Farm-servants' cottages and gardens are bottom left.

200–300 yards from the steading. This would reduce the risk of theft and fire (caused by the drunkenness of the farm servant or his wife) and keep mischief-making children away from the premises.[70] A site removed from the stink of the farm midden or dung hill was believed to be healthier too.

Sinclair provided a plan of a model layout for a farmhouse, steading and cottages (Plate 10). Two terraced rows of four cottages each are lined up alongside the road, within sight of the farmhouse (which is still backed onto the double dung heap!) but at a distance. The gardens are behind the cottages.

Many new farm cottages were erected beside the public road, where they would be out of the way of the steading and the fields (Plate 11). Farm cottages in Kincardineshire were 'regularly lined up' beside the road.[71] In Cleish, Kinrossshire, 'since the ground was enclosed, there are fewer cottars and their houses, instead of being placed irregularly in the fields, are built by the sides of the principal roads', in a row.[72] Terraced rows, sometimes double rows, took up less space and were popular with landowners and farmers, although aesthetically less pleasing. Today, terraces of farm workers' cottages are still found most commonly alongside country roads, sometimes beside the main farm track and always at a distance from the steading, which is usually set back some way from the road.

70 Sinclair, *Account of the System of Husbandry*, vol. II, 121.

71 Robertson, *GV Kincardineshire*, Edinburgh 1794, 417.

72 *NSA* Cleish, Fife, vol. IX, 44.

Gardens were usually enclosed, with the fence or dyke sometimes surrounding a row of cottage gardens, but commentators frequently remarked on the poor state of the fences. Patrick Neill criticised these gardens as 'ill-fenced' and called on landowners to build stone walls to keep out 'vermin'.[73] Constant invasions of the garden by deer or rabbits or by livestock and children destroyed crops and discouraged the gardeners.

Neill placed farm servants' kailyards (his term) at the bottom of his hierarchy of cottage gardens (see Chapter 3) – they have 'probably undergone no improvement for centuries'.[74] In early spring, he noted, cabbages, potatoes, leeks and onions were sown, and occasionally peas and beans. In east-coast counties gardeners grew 'German greens' and savoys and picked off the young outer leaves ('stowans') in April to make broth. Red curly kail and plain red kale were grown in Perthshire and the north of Scotland. There were no flowers, fruit bushes or medicinal herbs.[75] This was exactly the kailyard of old, with the addition of potatoes, fast becoming a vital staple of the poor.

But changes were happening. George Robertson recorded that

> when the farmers' own houses had arrived at their ultimate degree of refinement as at present [1820s, in the Lothians], those of their cottars or ploughmen were further improved also, more especially in their little gardens, which ... are also stocked with a greater variety of culinary store, as onions, leeks, and carrots, with a bush or two of gooseberries and currants, and even with flowers and flowering shrubs.[76]

Robertson claimed that at least in the Lothians the kailyard fed the servant's family all year round and there was, thanks to the potato crop, no lean time.[77] It should be noted that no other contemporary observer mentions flowers and fruit in farm servants' gardens at this time, so the practice was probably confined to the south-east, or even to parts of it.

Robertson was enthusiastic too about the kale grown in the kailyards of Midlothian: it was 'universally cultivated', needed less manure than cabbage, was enjoyed by cattle and even – possibly – enriched the soil.[78] The hinds of the south-east, regarded throughout the period as the cream of the agricultural workforce, were likely to have been ahead of the rest of the country in extending the range of their crops. But even here their 'first object is to plant cabbages' for the barley broth they mainly lived on. This was Berwickshire, where some servants also grew beans and 'white pease', but others had a small plot of onions,

73 Neill, *Gardens and Orchards*, 2.
74 Ibid., 9.
75 Ibid., 21.
76 Robertson, *Rural Recollections*, 87.
77 Ibid., 100.
78 Robertson, *GV Midlothian*, 11.

leeks and carrots. According to this observer, who greatly admired their efforts, 'the whole garden is constantly under crop' with two crops a year common; curly kale followed the lifting of the early potatoes in the summer. But no word of flowers, herbs or fruit.[79] Even in the 1860s there was a depressing lack of fruit in farm servants' gardens, put down to frequent flitting.[80]

While the absence of fruit and flowers can be explained by a reluctance to grow anything that could not be completely harvested by the term in May or was not portable, the absence of herbs is curious. Medicinal herbs would have been vital for the family. There are two possible explanations: they were too commonplace to be mentioned, although herbs in cottagers' gardens *are* mentioned in the *Statistical Accounts* and *General Views*; or the herbs they used were those that could be collected from the wild. One much later gardening writer though, Frances Jane Hope, does refer to southernwood (appleringie in Scotland) as one of the most popular medicinal herbs: 'In our ugly and utilitarian kale-yards there is almost always (and for it only) still room for a bush of Southernwood.'[81] Southernwood, which has a particularly and curiously pleasant smell, was carried to church by young girls to sweeten the musty atmosphere. It was also used as an insect repellent and was efficacious against childhood worms as well as weak stomach and feeble appetite.

Much later in the century the vegetables they were growing were still the old favourites: cabbages, savoys, curly greens (kale), carrots, onions and leeks, together with early potatoes.[82] Potatoes and cabbage still ruled supreme. The son of a shepherd in the Lammermuir hills recalled his parent's garden before the First World War. His father 'wasna much o' a gardener, he would plant potatoes, cabbages and so forth and that was aboot it. Didna go in for flowers much ... '. His father had enough additional ground to grow 'as many potatoes as you like'. There must also have been currant bushes, as he had memories of his mother making blackcurrant jam.[83] Currant and gooseberry bushes, grown for jam, were by now common in the south of Scotland, but less so further north, which surprised R. Hunter Pringle. Now that a larger proportion of the wage was paid in cash, wives could afford to buy jam from the van rather than spend time growing and picking and boiling their own fruit. Why the labourers of the south should have been keener fruit growers, or their wives keener jam makers, is not clear.

---

79 'Verus', 'On field gardening, and on the gardens of farm servants in Scotland', *GM*, 1826, 259–61, 261.

80 James Tait, 'Physiological distinctions in the condition of the Scotch peasantry', *THASS*, 1883, 1–20, 7.

81 Frances Jane Hope, *Notes and Thoughts on Garden and Woodland Written Chiefly for Amateurs*, London 1881, 78.

82 *RC Labour*, vol. III, pt. II, Fife, Kinross and Clackmannan, 61.

83 Lynn Jamieson and Claire Toynbee, *Country Bairns; Growing Up, 1900–1930*, Edinburgh 1992, 129.

In the Lothians, alongside the usual early potatoes, cabbage, onions, carrots and sometimes peas and beans, with savoys, greens and curly kail for winter use, Pringle noticed that some of the newer cottages not only had pot plants on the window sills but also flowers in the front garden and creepers on the walls. 'In summer many of the houses must be gay and bright with geraniums and roses.'[84]

In the north the *GVs* did not report on what servants grew, perhaps because the majority were unmarried. For later, it is difficult to find reports of farm servants here growing anything but potherbs and potatoes. An observer in the 1860s wrote that the gardens he saw in the north-east 'appeared to be employed solely for the growth of vegetables, and chiefly potatoes. I saw no signs of the pretty cottage flower gardens which are so common in England.'[85] This may have been down to the greater difficulty of gardening in the harsher climate, contentment with a more restricted diet of oats, herring and potatoes or simply lack of curiosity on the part of the reporters – Hunter Pringle was exceptional in this regard.

Many crops would have been grown from saved seed, but this practice has diminishing returns over time. Traffic between cottages and grand mansions enabled many poor gardeners to supplement their stocks and some farmers probably provided seeds too. Seeds and plants must also have been available to buy, from cadgers (pedlars) and markets in springtime. Linseed Friday in Kelso in the Borders took place in early April and was the day when country people came to town to buy their garden seeds.[86] The journalist Alexander Somerville recalled that as a young man in the 1830s he arrived one night at the home of Sandy Cowe, in the lee of the ruined castle of Innerwick. Sandy Cowe raised garden plants in the castle grounds and 'sent the seeds and plants throughout the country, to replenish the kailyards of the peaceful hinds – the ploughers of the farm fields of Lothian',[87] presumably via the cadgers. Later on, seed merchants (in every small town) and cheap collections by mail order as well as village shops and travelling vans would have made seeds available to gardening hinds.

HOW SUCCESSFUL WERE THE KAILYARD GARDENERS?

The 1970s farm-cottage garden in Fife was dug over entirely and planted with vegetables, mainly roots and brassicas, every May. In the nineteenth century the practice was similar: new servants, hired in March and given possession of the garden, then could be given time off by their old master to spend a Saturday in April at their new place, taking with them their seed potatoes and cabbage

84 *RC Labour*, vol. III, pt II, Edinburgh and Haddington, 113.

85 *RC Employment of Children, Young Persons and Women in Agriculture, 4th Report*, 1867, Appendix paragraph 29.

86 George Taylor, *From Kelso to Kalamazoo: the life and times of George Taylor 1803–1891*, ed. Margaret Jeary and Mark A. Mulhearn, Edinburgh 2009, 36.

87 Alexander Somerville, *Autobiography of a Working Man, by 'one who has whistled at the plough'*, London 1848, 365.

plants.[88] They then dug over the plot and, according to the Borders farmer Robert Shirra Gibb, if known and liked by their new fellow workers, would get help in the garden at dinner time (usually a two-hour break while the horses were rested).[89] A day to prepare the garden for the whole growing season, even with help, was not long, especially if the previous occupant had neglected it. But comments on how well or ill the farm servants tended their gardens are sparse. We also know little about what they looked like, possibly because they changed so little through the years and remained so utilitarian. The *NSA*, for example, has nothing to say about the state of these gardens: they are mentioned almost exclusively in connection with payment in kind.

The few assessments we have not only tell us about the farm servants' gardening practice but also apportion blame for the usually parlous state of the plots. Patrick Neill criticised servants for their neglect of their gardens in autumn and winter, when no digging was undertaken and the ground was weedy, strewn with decaying vegetation and infested with slugs and maggots. The result was that it meant the soil 'never works so kindly in the spring'. His sound advice was to get rid of perennial weeds, dig in autumn, apply manure and space plants more widely. However, he saw landowners as having the greater responsibility. The proprietor should make sure new cottages were south-facing, with their gardens in front, and should help to drain the ground where necessary. To ornament the gardens, and so that the cottagers could sell the fruit, the proprietors should plant apple and pear trees, and he lists reliable varieties. The laird, or his gardener, should demonstrate how to trench the plot, and should insist only on the keeping of ducks, which will eat slugs and caterpillars, as hens do too much damage.[90]

Sir John Sinclair saw things differently. He rated the value of farm servants' cows and potato grounds more highly than their gardens, partly because potato grounds helped to prepare waste or fallow fields for cereal crops. On the other hand, a small garden was 'of great benefit to the servant, but … a considerable loss to the farmer, from the dung it requires'.[91] He complained that farm servants did not grow enough of 'the onion tribe', even though they ate plenty. However, he agreed with Neill on the siting of the gardens – they should be in front of the cottage and facing south.[92]

The Berwickshire correspondent of the *Gardener's Magazine*, quoted above, had only compliments for the gardening skill of the farm servants he knew. The 'kitchen-garden is of immense benefit to a Scottish family, especially when the owner has a taste for dressing it sufficiently, of which the farm servants

88 H. Stephen, *The Book of The Farm*, Edinburgh 1844, vol. II, 52.
89 Robert Shirra Gibb, *A Farmer's Fifty Years in Lauderdale*, Edinburgh 1927, 15.
90 Neill, *Gardens and Orchards*, 27.
91 Sinclair, *Account of the System of Husbandry*, vol. II 131.
92 Sinclair, *General Report of the Agricultural State and Political Circumstances of Scotland*, Appendix vol. I, chapter III, appendix no 4.

in Scotland are seldom destitute'.[93] But other commentators were scathing, blaming the landowners and farmers for the state of the farm cottages, on the one hand, and the ignorance of their occupants on the other. A (self-described) 'intelligent East Lothian farmer', taking the train in the summer of 1847 on the new Edinburgh to London line, occupied his time by looking out at the passing fields and cottages. When he got back home he wrote to *The Scotsman* with his impressions of the differences between the Scottish and English countrysides. English farming in every respect was superior to Scottish – productivity, crops and cattle were all more advanced and more scientifically based. English cottages, wreathed in roses, were more attractive too, and he urged Scots landowners to replace the 'barrack rows of farm servants cottages', which were a disgrace to landowners and farmers alike. 'Let them be detached, and then we will see the flowers growing and neatness of their gardens.'[94]

The ploughman James Alexander scolded his fellow servants for ignorance, while at the same time blaming farmers for their indifference to the social and moral condition of their workforce. As a result 'we not infrequently see the "kailyard" lying half-waste – many useful vegetables of easy cultivation being altogether neglected, because the proper manner of using them is unknown.'[95]

In the north of Scotland the continuing housing shortage would not have encouraged gardening. Farm workers unable to find cottages when they married and forced to keep their families in villages and towns rarely had gardens, or, if they did, they were too small to be of practical use in feeding the family. By the 1890s gardens here were 'devoted to potatoes' and not well looked after. Constant migration of farm servants from farm to farm was held to blame.[96]

In the more advanced and wealthier counties from Perth and Fife down to the Borders 'it is not unusual now to add in front a small plot for flowers'.[97] An essay in a *THASS* series on agriculture county by county in the 1870s complimented what were probably farm workers in Midlothian and West Lothian:

> A plot of garden ground is also attached and the thrift of the cottager is seen in the fine crops of vegetables which adorn his kitchen allotment. In fact, an air of quiet comfort everywhere seems to pervade the working classes, and in no section of Lothian agriculture has there been more melioration during the past quarter of a century than in the condition of these honest tillers of the soil.[98]

---

93 'Verus', 'On field gardening', 261.

94 Anonymous letter in *The Scotsman*, 14th July 1847, 3.

95 Alexander, 'Present condition of farm servants', 15.

96 *RC Labour*, vol. III, pt I, Moray, Banff and Nairn, 112.

97 *RC on Employment of Children, Young Persons and Women*, Appendix, 78.

98 Thomas Farrall, 'On the agriculture of the counties of Edinburgh and Linlithgow', *THASS*, 1877, 1–66, 52.

For the Royal Commission on Labour of 1893, twenty-five farmers, mostly running arable farms in the Lothians, were asked about their servants' gardens. Their responses were generally positive, if lukewarm: 'carefully tilled and well kept', 'well tilled and fully utilised' and 'as a rule well cared' were the most common remarks. In fact only one respondent, who farmed near the mining town of Tranent, was negative ('not well cared'), but he seems to have had a poor opinion of his workers generally, describing them as 'unsettled and listen[ing] to agitators about the coalpits'.[99] Generally, gardens were given over half to potatoes and half to vegetables and the crops rotated from year to year; some divided their plot into three, a third each for potatoes, vegetables and fallow. But blight was common, caused it seemed by using less resistant seed in order to raise an early crop, by use of the same seed year after year and by not taking enough care with rotations – in other words, by poor (and ignorant) gardening practice.

Glances over the garden wall and comments by farmers, who were notoriously bad gardeners, do not tell us much, but, if there were some bright spots about the farm cottages, the general picture was depressing. R. Hunter Pringle, one of the most conscientious, thorough and reliable of the contributors to government inquiries – and a farmer himself – held extensive discussions with workers in central and southern counties and concluded that many workers felt that the work of 'delving' was a 'botheration' and not worth the effort. Sometimes half the garden might be left undug and, where it was dug, this would be done in the spring, not the autumn – Neill's old complaint. Pringle also seems to have understood why gardening had become such a burden:

> There are no idle hands, no lads hanging about who could see to the digging and cultivation of a large garden [because they were working elsewhere, or at school]. The labourer himself returns from the field a tired man, indisposed at any time of the year to tackle the tilling, planting or weeding of a large garden.[100]

GARDEN SIZE

The size of the garden was critical. Too small and it would not feed the family; too large and either the servant would neglect his duties on the farm or he would be able to cultivate only part of it. The evidence we have on this important question is both patchy and contradictory. The early farm servants' gardens seem to have been small. Robertson says only two or three falls (seventy-six to 114 square yards approximately) was the usual size in Midlothian.[101] According

99 *RC Labour*, vol. III, pt II, Edinburgh and Haddington, appendix, 145.

100 Ibid., 61.

101 Robertson, *GV Midlothian*, 166.

to Patrick Neill (who does not give the sources of his information), farm servants' gardens across Scotland varied widely in size, between three and eighteen falls (103 and 686 square yards), and he argued that even eighteen falls was insufficient to keep a family of six in potatoes during the winter unless they also had a potato rig in the farmer's field. Such a small plot meant that crops were planted too close so they became overcrowded, resulting in, inevitably, poor yields. Continuous cropping with greens and cabbages, to ensure a constant supply of potherbs, exhausted the soil.[102]

Neill advised landowners to allow twenty-four falls (half an acre) per household, with a pig. More would be useless and even harmful, as it would take up too much time, exhausting the worker (and so injuring the farmer too), and need more manure than could be provided by the farm.[103]

A little later, contemporary comments suggest that, in some places at least, gardens became larger and adequate for feeding the hind's family. Robertson, a reliable and authoritative witness, noted in the 1820s that the servants' gardens he knew in the Lothians in the 1820s were larger than they had been.[104] However, in Markinch in Fife in the 1830s gardens were only eight falls[105] and several NSA accounts describe hinds' gardens as small.

Farm servants' gardens in the last few decades of the nineteenth century got smaller. A government inquiry in 1867 reported that in the counties between Perthshire and the Borders 'gardens of from ten to twenty poles (one sixteenth to one eighth of an acre) are always attached to farm cottages.'[106] None of the witnesses recorded in this Report appear to have given garden measurements, so we cannot tell where Mr Colley found this figure. Even one eighth of an acre is small – a quarter of what Patrick Neill recommended. Twenty years later, another inquiry heard that gardens in the north were around one sixteenth of an acre.[107] 'Quite large enough for a working man's garden' commented the witness, the Superintendent of the Poor for nine northern counties. 'That is not our experience in England', was the rather acid response of the chairman, the radical MP Charles Dilke. Further south garden dimensions varied, it seems, from farm to farm. Mr William Smith of Brechin, a leading Angus farmer, told the inquiry that his farm servants' gardens were nearly one sixth of an acre – 'as much of a garden as they can well manage' in the summer evenings and in their

[102] Archibald Gorrie, 'Observations on the cottage gardens in the Carse of Gowrie', in Patrick Neill, *On Gardens and Orchards*, Appendix D, Edinburgh 1813.

[103] Neill, *Gardens and Orchards*, 21.

[104] Robertson, *Rural Recollections*, 87.

[105] *NSA* Markinch, vol. IX, 671.

[106] *RC Employment of Children, Young People and Women*, 78.

[107] *Report of HM Commissioners for Inquiring into the Housing of the Working Classes, 2nd Report: Scotland*, 1884–5, 97.

spare time.[108] Robert Duncan, secretary of the Scottish Ploughmen's Union,[109] which was based in Perthshire, told another inquiry a few years later that, to his knowledge, gardens were twelve to sixteen square yards, which would have been impossibly small.[110]

In the central counties of Fife, Clackmannanshire and Kinross in the 1890s we find Pringle again, this time questioning farmers about the size of their workers' gardens: the answer was between one eighth and one twentieth of an acre, planted with the usual potatoes and 'plain vegetables', often poorly cultivated. Larger gardens, they said – up to a quarter of an acre – had been more common in the 1870s.[111] In West Lothian gardens were between 100 and 300 square yards – not very different from Fife.[112]

Large farms were taking on more men in the epoch of high farming, but do not seem to have provided much more land for their gardens. Fenton Barns in East Lothian was renowned as a model farm in the mid-nineteenth century, in the most advanced agricultural district in Britain, and the tenant, George Hope (1811–1876), was an internationally famous agriculturist. In the early 1850s up to nine farm-worker households shared a fairly large piece of garden ground between the well and the pump.[113] By the 1890s ten cottages had been added, each with a long narrow garden behind, apparently occupying the same area as the earlier gardens.[114]

Statistics collected in 1886 give some idea of the variation in garden size between counties at this time. The *Return of Allotments Detached from and Attached to Cottages, in England and Wales, and Scotland, 1886* was not interested in small plots and gives data only on attached farm workers' gardens of one eighth of an acre and over, so is not particularly helpful in relation to Scotland, where many farm workers' cottage gardens were, as we have seen, smaller than this. Of the seventeen counties in the east of Scotland, in seven there were no attached gardens of one eighth of an acre or larger. In six there were fewer than 500 of this size and in only three – Roxburghshire, Fife and Perthshire – were there more than 500. There were nearly 2,000 in Fife (why is unknown).[115] Along the Moray

108 *Report into the Housing of the Working Classes*, 98.

109 William Smith was, incidentally, a strong opponent of small-holdings, and of the Scottish Ploughmen's Union, which campaigned for them (*Dundee Courier*, 11th March 1889, 3).

110 *Report from the Select Committee on Small Holdings*, 1889, 484.

111 *RC Labour*, vol. III, pt II, Fife, Kinross and Clackmannan, 60.

112 *RC Labour*, vol. III, pt I, North Lanarkshire and Linlithgowshire, 86.

113 Haddingtonshire, Sheet 5 Ordnance Survey (6 inches to the mile) 1st edn 1854, available at https://maps.nls.uk/view/74426938.

114 Haddingtonshire Sheet V. NW, Ordnance Survey (6 inches to the mile) 2nd edn 1895, available at https://maps.nls.uk/view/75632016.

115 *Return of Allotments Detached from and Attached to Cottages, in England and Wales, and Scotland*, 1886, Scotland, Table II.

Firth (in the counties of Moray, Nairn and Banff) in the 1890s most farm cottages had attached gardens of around one eighth of an acre, but there were reports of cottages without gardens too. 'All seem to agree that, while he is in service, a farm servant cannot look after a croft, or anything more than the little patch of garden ground, perhaps an 1/8 of an acre, which he has.'[116]

Not a single worker amongst those giving evidence to this Royal Commission complained that their garden was too small. Several farmers declared that they had offered extra ground, and the manure to dung it with, but these offers were never taken up. The garden had become less of a necessity now that a growing proportion of the ploughman's wage was in cash, much of which went on purchases of food from village shops or grocers' vans.

WORK FOR THE FAMILY

Throughout the period we are dealing with work on the farm was presumed to be a family effort. Sir John Sinclair had pointed out the obvious advantage of married over unmarried servants: 'their wives, and their children, from the moment they can handle a hoe or a weed hook, except in the dead of winter, are never short of *out-work*, for which they receive ample wages.'[117] Patrick Neill had also urged farmers to hire 'tried and trusty *married* servants' (Neill's emphasis) who will bring with them their wives and children to work in the fields at weeding and hoeing.[118] There is evidence that women worked hard in their own gardens too. In the 1890s, 'a smart wife will contrive to have some bright flowers.'[119] 'There was a flower border which Mother used to look after', recalled the shepherd's son. Gardens that were neat and well-cultivated were often acknowledged to be the achievement of the farm worker's wife. 'Whenever I heard of a man whose garden was well kept and productive, it often turned out to be the wife who did the work', wrote R. Hunter Pringle. One summer evening he visited a garden in Kinross-shire, 'nicely laid out and evidently well-cared for', which was kept by the 'active and thrifty young wife' of a ploughman. She was hoeing the rows of vegetables, helped on this occasion by her husband, who confessed, shamefacedly, that he did this for company, and because 'he didna like to be doing nothing when she wis working'.[120]

Children in the early part of the century were mainly a menace, akin to marauding livestock and vermin: they stole crops and trampled the ground. Expectations that they would help in the garden – and so become accustomed to farm work – may have been more hopeful than realised. Compulsory schooling,

116 *RC Labour*, vol. III, pt I, Moray, Banff and Nairn, 112.

117 Sinclair, *Account of the System of Husbandry*, 260.

118 Neill, *Gardens and Orchards*, 27.

119 *RC Labour*, vol. III, pt I, Moray, Banff and Nairn, 112.

120 *RC Labour*, vol. III, pt II, Fife, Kinross and Clackmannan, 61.

brought in in 1872 (seven years earlier than in England) must have reduced children's participation in the garden, as well as depriving them of paid work for most of the year.

PIGS, POULTRY AND PRIVIES

It was not only the crops raised in the garden that gave it value. It might house pigs and poultry too – some kept for their commercial value, some for the family's own use. Pig-keeping at the beginning of the 1800s seems to have been more common near towns, as near Berwick, where 'every one who has an opportunity, has a pig either tethered at the door, often on highways, or grunting in the house', fed, at this time, on waste food and milk.[121] The increasing popularity of potatoes helped pig-rearing to spread, but the attitude, hostile or otherwise, of the master was a determining factor, as well as proximity to a market. For the farm worker's family pigs were the most valuable domestic animal they could raise. Bought as weaners in the spring, fed on kitchen scraps and potatoes, slaughtered in the autumn or winter and then usually sold, they could make a vital contribution to the family's cash resources and, of course, their dung was essential to the garden. In the 1890s in Fife there was a sty at every cottage on most farms, the pig kept for the family only.[122] In the Lothians rules were more relaxed and servants could keep several pigs for their own use and for sale.[123] Farmers in the north on the other hand discouraged the practice, fearing the theft of fodder from byre and stable.[124] However, where pigs were kept, it was noted that the gardens got more attention.

Poultry-keeping was less common and also partly determined by distance from a market. In Midlothian, in the early years of the nineteenth century, all farm servants kept poultry and sold the eggs in Edinburgh;[125] Berwickshire eggs were exported to London from Berwick-on-Tweed. 'Barnyard fowls' were kept by almost every cottager in Aberdeenshire, and both hens and ducks were common in the Mearns.[126] But elsewhere they were less popular. Poultry could only be fed on grain, not on kitchen scraps and potatoes, and needed to be locked up at night. And they scratched up crops and gardens too. Like the pigs, they were often discouraged by the farmer as a menace to his grain store. By the end of the century poultry-keeping seems to have been confined to the north and to Fife, while in the Lothians farm servants were simply not allowed to keep poultry.

---

121 Lowe, *GV Berwickshire*, 23.

122 *RC Labour*, vol. III, pt II, Fife, Kinross and Clackmannan, 62.

123 Ibid., 113.

124 *RC Labour*, vol. III, pt I, Moray, Banff and Nairn, 112.

125 Robertson, *GV Midlothian*, 166.

126 Skene Keith, *GV Aberdeenshire*, 506; Robertson, *GV Kincardineshire*, 112.

And, amongst the pigsties, henhouses, privies, ashpits and perhaps washhouses, later gardens would also have had to make room for drying greens.

On the eve of the First World War, farm servants' gardens, if well-kept and displaying flowers and fruit, would have been indistinguishable from cottage gardens. We cannot tell how many were. But not many of the gardens played the vital role they once had done. Smaller and probably less well-cultivated, they mostly produced the same humdrum range of vegetables as a hundred years before. The households they had to feed were smaller, while the gradual rise of money wages and the arrival of the grocers' carts meant more, and more varied, food could be bought, so home-grown vegetables became less essential to the farm workers' family's standard of living.

A study of the budgets of forty-two rural labourers in England by B. Seebohm Rowntree and M. Kendall pointed out that 'less than a twelfth of the food consumed by these families was derived from their gardens. And it must be remembered that from a garden of the usual size produce cannot be obtained the whole year round.'[127] There is no comparable study for Scotland at this time, but it is likely that the shrunken farm workers' gardens, not always fully cultivated as we have seen, produced even less, even if they were able to raise kale and leeks and kept a store of potatoes through the winter.

For those with little interest in gardening, the decline of the garden must have come as a relief, and meant that spare time, still at a premium, could be spent on other, more enjoyable and perhaps more stimulating activities. If a town or village was within reach, it could now offer more entertainment[128] than the garden. But many, living far from the village, still on lower cash wages than other workers and having so little free time outwith the winter months, were largely excluded from the social life that revolved round the church, the pub, the shop, the smithy and the school.

But, as in the earlier decades, what most discouraged work in the garden were long hours and long weeks, harsh weather, arduous physical work and frequent flitting. There was little encouragement to keep the garden well cultivated – for example, there was no compensation for any crops left in the ground when the house was given up, or for any manure spread to improve the soil.[129] Farm workers' nearest neighbours were their masters, and most farmers showed little interest in gardening. Contemporary writers, even other farmers, found this inexplicable. George Robertson noted in 1813 that Kincardineshire farmers'

127 B. Seebohm Rowntree and M. Kendall, *How the Labourer Lives, a study of the rural labour problem*, London 1913, 307.

128 In the 1880s Fochabers, a village in Moray, held dances, concerts with English singers, cycling races and public lectures, and it had a reading room (John Tully, *A half-century glance backwards: a paper read in Fochabers Mutual Improvement Association on February 12, 1930*, Banff 1930, 15).

129 *RC Labour*, vol. III, pt I, Moray, Banff and Nairn, 119, 124.

gardens were often so neglected that they did not even grow potherbs.[130] In general, they were 'unsightly wildernesses'.[131] Farmers were even accused of boasting about their ignorance of gardening.[132] As a result, 'while he keeps his garden and shrubbery and little avenue [for farmhouse gardens had come a long way] in a slovenly manner, it is not to be expected in servants to evince a desire to excel their master', scolded Henry Stephen, the author of the hugely popular *Book of the Farm*.[133] In spite of this, many farm workers and their wives did take pride in their gardens. A few have been found entering cottage garden flower shows, although almost exclusively in the traditional vegetable classes. East Lothian hinds seem to have been most likely to enjoy gardening, although just as likely to be addicted to frequent flitting as elsewhere.

The general opinion, however, was that farm servants were not particularly interested in gardening, beyond growing potatoes and a few other vegetables. 'Little effort is spent in trying to make the gardens add to the amenity of the cottages', noted another government inquiry, 'although occasionally effort may be made at the cultivation of flowers'. It was admitted – once again – that, given their 'hours of labour and the nature of their daily work, it is unreasonable to expect that [gardening] should be [popular]'.[134] A monthly publication, *The Plough*, launched in the 1890s by Robert Dempster, ploughman, stone quarrier and the man who founded the Ploughmen's Union, did include gardening advice, but, as the periodical lasted only two years, this may have had little impact.[135]

R. Hunter Pringle can have the last word. His conclusion to his report on the farm servants of the Lothians recalls the moralising tone of the early commentators, but also puts in a plea for education for farm servants as a way to make gardening more productive and more pleasurable:

> Their knowledge of cottage gardening is very limited and imperfect. Although I saw many well-kept gardens, still there was a complete absence of what I might call 'intensified' management. The rotation of crops now in vogue is too narrow. Many useful and suitable herbs and vegetables are ignored. Floriculture is not understood. Now, the Lothian labourer is naturally fond of flowers and gardening, and it would be a great matter if he were taught how to make the most out of his little garden. ... it is planted with a few crops which come under the general term 'greens', but the system of 'pull a cabbage, plant a cabbage' is never followed. Consequently, a garden does not return

130 Robertson, *GV Kincardineshire*, 299.

131 Menteath, 'Cottagers' garden societies', 783.

132 David Gorrie, 'Hints on the Improvement of Cottages', *Journal of Agriculture*, 1848, 1–27, 1.

133 Stephen, *Book of the Farm*, vol. 2, 435.

134 *RC Housing of the Industrial Population of Scotland*, 170.

135 Mary Dempster Hewitt, 'Robert Dempster: a forgotten activist', *Scottish Local History*, vol. 88, 2014, 3–8, 7.

its full value to its owner. Technical education has been designed to meet this case, and there can be little reason to doubt that a practical instructor would be welcomed in these localities. To me, it would appear that a thorough knowledge of cottage gardening, and for the women, cottage economy, would tend to settle the labourers and mitigate the many evils consequent on the flitting mania. When a man and wife appreciate the pleasures of a good house and pretty and productive garden, and have no faults with their employer or employment, they are less likely to shift their quarters than when both house and garden are in a hugger-mugger confusion.[136]

136 *RC Labour*, vol. III, pt II, Edinburgh and Haddington, 122.

CHAPTER 3

COTTAGERS' GARDENS

> Every cottage had its conveniently situate garden ...1/8 of an acre. Early potatoes were its staple product, and as these were cleared off, the land was planted with Cabbages, but chiefly hardy winter Greens; but there were always plots of summer Cabbage, Leeks, Potato, and Onions, sometimes a patch of Chamomile, Hyssop, and other simples, and perennial flowers, and sometimes several hives of bees.[1]

So wrote a correspondent to the leading national gardening magazine, the *Gardener's Magazine*, recalling the labourers' cottage gardens of East Lothian of the 1840s. To find out how typical these were, then and later, how they came about and why they were different from the kailyards is the purpose of this chapter.

Cottagers' gardens as we describe them here were a by-product of the massive social and economic changes that began in the last half of the eighteenth century. 'Not until about the end of the 18th century did cottage gardening with its combined vegetable and flower culture make headway in Scotland.'[2] Key to the change was the demand by improving landowners and farmers for labourers for draining, enclosing, building, road-making and making and maintaining farm equipment. Like the farm servants, they needed to be housed within reach of the farms, and every house came with a garden. The generosity of landowners and farmers in handing over land often initially rent-free is explained by this demand for labour. The land for gardens allowed to other workers, from miners and mill workers to lighthouse keepers, served the same basic function of providing food for workers and their families far from villages. As well as these groups, there were the village gardeners, skilled tradesmen such as masons, joiners, smiths, saddlers, wrights, tailors and shoemakers and handloom weavers. These workers flourished under improvement, as their specialist skills were required for new steadings, new farmhouses, ploughs, carts, harnesses and so on, while rising real wages for many created a market for clothing and household goods.

1 *Gardeners' Chronicle*, 31st March 1866, 332.

2 J. A. Symon, *Scottish Farming Past and Present*, Edinburgh 1959, 415.

Not all lived in cottages – some mill workers lived in tenements – but all had gardens attached to their homes to supply them with food for their families. Was the contribution of these gardens to their standard of living similar to the case of the farm servants, or did they also serve other functions? What was the reason for the wider variety of crops in these gardens? Did they have significance for them other than simply filling the kailpot?

More free time, eventually, greater security of tenure[3] and the stimulation and encouragement of living in a community – all these distinguished cottagers' lives from farm servants',[4] although their gardens might be much the same size.[5] Many cottagers had more time for gardening than farm servants. Spare time was not always welcome: day-labourers and handloom weavers on piecework were often out of work; village tradesmen were also unlikely to be employed full-time all year. Miners and mill workers, on the other hand, worked very long hours in very unpleasant and unhealthy conditions but in spite, or perhaps because, of this we find gardeners, including very keen gardeners, in both these groups.

Amongst others, James Headrick urged landowners to build labourers' cottages: where the houses were held from the farmer they 'only possess their houses from year to year, hence any little improvement they may make only subjects them more to the will of their master'.[6] Landowners did, gradually, start to take over responsibility for labourers' housing and so give more security of tenure.

Cottagers, usually tradesmen, who built their own homes or paid rent or a feu were more secure. Several parish reports in the *OSA* mention such arrangements: in Little Dunkeld in Perthshire[7] a house and small garden could be rented for between 10s. and 16s. Cottars who held their house and garden in return for

3 In the earlier part of the period there was often little security of tenure for the labourers occupying cottages for rent or in return for farm labour, a growing practice that suited improving farmers. Labourers were at the mercy of the farmer's need, or whim, and eviction or a rent rise at the year's end were constant threats. However, frequent flitting was not new: it had been customary for cottars to move from fermtoun to fermtoun, perhaps not very far, sometimes accompanying the tenants for whom they worked (this could be required by the lease), sometimes because of bad harvests. Younger sons were usually obliged to move out and seek work elsewhere (Whyte, 'Population mobility in early modern Scotland', in R. A. Houston and I. D. Whyte (eds), *Scottish Society 1500–1800*, Cambridge 1989).

4 The vital importance of a house and garden is illustrated by the rule in Redgorton in Perthshire: 'When the young folks of either sex, are suspected of too much intimacy, the young man is called before his master; and if he refuses to marry the lass he professes to love, he is immediately dismissed from the work; but if he marries, which is generally the case, he gets a house and garden, and a small compliment' (*OSA* Redgorton, Perthshire, vol. XV, 540).

5 Devine, 'Social stability', 61.

6 James Headrick, *GV Angus or Forfarshire*, Edinburgh 1813, 137.

7 *OSA* Little Dunkeld, Perthshire, vol. VI, 373.

services on the farm might also rent an acre and a cow's grass, sometimes paying very dearly – 30s. for an acre of arable land and a cow's grass in this parish.

Again unlike the farm servants, cottagers lived in villages, hamlets and little huddles of houses that sprang up unofficially on the edge of moors. The history of cottage-garden societies and best-kept garden competitions is among the best evidence we have so far (in the absence of a history of the Scottish lowland village) of the power of community to influence and shape the garden. The rise and success of the cottage garden societies testify to the growing enthusiasm for raising good-quality fruit, flowers and vegetables. Many workers, perhaps especially those, such as the weavers, whose work kept them inside all day, would rise at dawn to 'dress' their plots before work and toil afterwards until late in the evening.

TIME FOR GARDENING

Sufficient time for gardening was essential if the garden was to feed the family. The few hints there are suggest that the cottager tended his garden early in the morning and late into the evening, in the summer months at least, or was expected to. The horticulturist Archibald Gorrie, however, calculated that a cottager needed to spend twenty-six days a year, and his children six days, in order to bring in a harvest to feed a family of six.[8] Spring, early summer and October would be the busiest months. Was it feasible for a full-time worker to manage this? Long working hours, Scottish weather and church-going habits make it unlikely. The assumption seems to be that the labourer is not employed full time.

Until the 1870s there are few signs of over-supply of labour and so of underemployment. During harvest, the busiest time of the farming year, migrant workers from the Highlands and Ireland were on every farm. In the north-east 'there is little evidence of any large body of men lacking employment and staying in the countryside':[9] they had the options of moving to the towns and cities or emigrating. This was even more the case after the agricultural depression cut demand for less skilled labour. A high proportion of farm workers in Scotland were women, all paid by the day or the week, and this also suggests that there was plenty of work; women were paid around one-third of the men's rate.

In the first half of the century handloom weavers on piecework or working as 'customer weavers' could rarely expect to be employed full-time, especially the many less skilled workers who wove the coarser linens. Gardening and even smallholdings fitted in well. The *GV* for Angus records that 'As their trade admits without prejudice of short interruptions they cultivate these little farms very

---

8 Archibald Gorrie, 'An essay on cottage gardens', in J. C. Loudon, *A Manual of Cottage Gardens, Husbandry and Architecture*, London 1830, 62–72, 69.

9 Malcolm Gray, 'Farm workers in North-East Scotland', in T. M. Devine (ed.), *Farm Servants and Labour in Lowland Scotland 1770–1914*, Edinburgh 1984, 21.

successfully, greatly to the benefit of their own health and the improvement and beauty of the country.'[10] Not all weavers' employers regarded the short interruptions so benignly and often complained at the delays in production it caused.[11] Weavers also interspersed their stints at the loom with field labour, especially during harvest – sometimes treated as a holiday, when weavers from the towns would join in.

At this time, the memoirs of worker gardeners testify to the variety of non-farming labouring jobs available to strong, young – and unmarried – men. From one harvest to the next in the mid-1820s Alexander Somerville had jobs digging drains, in a lime kiln, hedging and ditching and gardening, as well as ploughing – apparently without a break. When unable to get a hire as a servant he headed for Edinburgh, where his brother found work for him in his saw-pit.[12] Cottagers with more time for their gardens would have been older and less mobile – and more likely to be married or living with family members – so perhaps less likely to expect continuous employment. The cottage gardeners who won prizes at flower shows or in the best-kept garden competitions and who can be identified through the census records did tend to be older – over forty, sometimes much older. Prizewinners, of course, would have been the keenest gardeners.

THE EARLY COTTAGE GARDENS

One of the earliest references to cottage gardens appears in Lord Kames' *The Gentleman Farmer*, published in 1776, and one of the most popular books on improvement amongst progressive landowners. Kames boasted: 'my labourers have good kitchen-gardens, where onions, leeks, cabbage, turnip, and potatoes are sown in drills.'[13] Still, there were a few critical voices:

> The condition of the people, for the most part, is rather more than tolerable, and they are apparently contented with it. It might, however, be, in some measure, improved at no great expense, by making their cottages more comfortable and convenient, by raising better fences to their gardens, and introducing among them to culture of a few more nourishing vegetables.[14]

Patrick Neill's 1813 report *On the Gardens and Orchards of Scotland* divided cottage gardens into four classes: feuars'; village gardens; gardens of day-labourers and the humbler tradesmen in informal settlements or hamlets; and

[10] Headrick, *GV Angus*, 126.

[11] Catherine Rice, *'All their Good Friends and Neighbours': the story of a vanished hamlet in Angus*, Dundee 2014, 36.

[12] A. Somerville, *Autobiography*, 100–4.

[13] Henry Home, Lord Kames, *The Gentleman Farmer: being an attempt to improve agriculture by subjecting it to the test of rational principles*, Edinburgh 1776, 398.

[14] *OSA* Tealing, Angus, vol. IV, 102.

the farm servants' kailyards. In general, he found that the poorer the gardener the worse the gardening. Feuars were better-off tradesmen who held a house and garden from the landowner together with an acre or two of arable land in return for an annual payment, for land released by landowners usually as a way of creating or expanding villages. According to Neill this group had the largest gardens – from eight to twenty-four falls – and the best cultivated. They grew a wide variety of fruit and vegetables, but Neill makes no mention of flowers.[15]

Neill's village gardeners, who seem to have been artisans and 'manufacturers' (handloom weavers), benefited above all from their middle-class neighbours, whose presence, he believed, stimulated imitation and competition. Neill was not, like some commentators, obsessed with village drunkenness, but he did notice that these gardeners did not frequent the alehouse: 'indeed it is observable, that in those places where cottagers have tolerable gardens, a taste for keeping them well, is prevalent, and few instances of dissipation occur.'[16] Cottagers valued their garden at a quarter and a half of the rent (or 15s. to £2 8s. 'according to circumstances') and might choose a cottage just for its garden, measuring an eighth to a quarter of an acre. A single reference in the *OSA* claims that the value of the garden produce to a labourer's family was the equivalent of the rent – and that although they had to pay 10s. for fuel this was completely recouped by the potato crop, dunged with the ashes.[17]

Neill claimed that where the soil was good and the garden large enough some 'may be found who consider themselves as paying their rent for their gardens only', the houses being regarded as a 'mere appendage'.[18] On the other hand, Neill blamed the annual leases labourers had to accept for the poor state of so many of their gardens. What was the point of improving the soil if the garden was to be lost to another after a year? However, he was also critical of those who neglected their plots: cottagers were often either too ignorant or too idle to undertake winter digging, for example, and tended to overcrowd their plots with plants. While rotations were possible in larger gardens, many were much too small – ten falls – and so were cropped year after year with greens. Even on

---

15 Ibid., 20–1.

16 Ibid., 22.

17 *OSA* Caputh, Perthshire, vol. IX, 501. Not nearly enough data has been found to allow an assessment of the value of cottage gardens to their occupiers either at the start of the nineteenth century or later. In England, allotments (much larger than the average Scottish labourer's garden) were estimated to add one fifth to the wage; less – £1–£1 10s. per annum in the 1860s; and ten to twenty per cent by the start of the First World War (John Burnett, 'The English village labourer revisited. Self-provisioning and the commercializing of diet', in M. Hietala and T. Vahkatari (eds), *The Landscape of Food*, Helsinki 2003, 89).

18 Neill, *Gardens and Orchards*, 19.

the excellent black loam of the Carse of Gowrie, complained Archibald Gorrie, this exhausted the soil.[19]

Neill's *Report* does not differentiate between regions. From the *Old Statistical Account* and the *General Views* it seems that better gardens followed in the wake of the march of 'improvement' across the Lowlands. Both these Sinclair projects were commissioned during the mainly prosperous years of the French and Napoleonic Wars, when agriculture was booming, labour in demand and wages rising, and their purpose was in large part to record and promote improvement. While the *OSA* authors, mostly parish ministers, were usually more interested in the gardens of the landowners, and their own, the contributors to the *GVs*, often farmers, were more careful to take note of flourishing cottagers' gardens where they existed.

Nevertheless, even in the *OSA* there is the odd comment: both encouraging – 'the gardens produce every necessary vegetable and root, together with all kinds of common fruits'[20] – and enraged – 'There is scarcely a garden in the parish that deserves the name; culinary productions being more regarded than ornament or elegance.'[21] For Lowland Perthshire Robert Thomas, clearly a gardening enthusiast, produced two parish reports for the *OSA* that extol the gardens of rural workers. In Forgandenny he found a great range of vegetables, gooseberry bushes (which sheltered the bee hives), herbs 'and some flowers, which serve both for ornament and use' grown in the gardens of the 'common people'.[22] In Scone, just outside Perth, he notes:

> the handicraftsmen [handloom weavers], at their leisure hours, ..., begin to pay more attention to their gardens than formerly; a certain indication of the thriving state of this part of the country, men commonly attending, first, to what is necessary, and then to what is commodious and ornamental.[23]

On the other hand, also in Perthshire, 'no attention is paid to the advantage of a kitchen garden' in the largely unimproved parish of Kilmadock, where many were extremely poor, their houses collapsing 'wretched huts' and the roads impassable.[24]

GV reports often provided more detail. Cottage gardeners in the Borders and the Lothians, already in the forefront of improvement, won praise for good

19 A. Gorrie, 'Observations on the cottage gardens', 191.

20 *OSA* Auchtermuchty, Fife, vol. VI, 343.

21 *OSA* Coupar Angus, Perthshire, vol. XVII, 5–6.

22 *OSA* Forgandenny, Perthshire, vol. III, 303–4. Forgandenny kept its charm: 'when the woods are in full foliage, and the gardens gay with flowers, it would be difficult to find a sweeter spot' (Thomas Hunter, *Woods, Forests and Estates of Perthshire*, Perth 1883, 154).

23 *OSA* Scone, Perthshire, vol. XI, 76.

24 *OSA* Kilmadock, Perthshire, vol. XX, 52.

gardening.[25] Many of the gardens in Selkirk and Galashiels were 'neatly dressed, yielded a profusion of useful vegetables, and were ornamented with shrubs, flowers, and bushes, bearing the smaller fruits'.[26] In Roxburghshire labourers had kailyards, growing only potatoes with a few cabbages and some open kail for winter. 'Of late, a few beans and turnips have been introduced but they are greedily devoured by children before they are fit for use'.[27] The Roxburghshire tradesmen, on the other hand, 'have neat and curious gardens carefully and skilfully managed, producing not only the best kinds of common vegetables but many uncommon plants and flowers, both for beauty and for use'. Amongst these last were rhubarb, teasels, woad, moss roses and yellow roses.[28] At this time, rhubarb was grown for its medicinal properties only; teasels were mainly grown for combing (carding) wool before spinning; woad was used in dyeing. These gardens thus had an 'industrial' function, reflecting the gardeners' and their families' occupations.

In Angus,

> every cottager and country tradesman has also his kail-yard, or inclosed plot of ground, of more or less extent, in which he raises cabbages, coleworts, leeks, onions, and all sorts of esculent [edible] roots. The gardens of the cottagers and tradesmen are generally much better attended to than those of the farmers; and many of them are stocked, not only with gooseberries and currants, but with fruit trees, and adorned with flowers.[29]

Further north, improvement was in full swing in Kincardineshire. The farmer George Robertson greatly admired the gardens attached to the scattered cottages or those in the little settlements created by weavers, tailors, shoemakers and other tradesmen in out of the way places, on waste or marginal land. While farmhouses here were often to be found on top of a windswept hill or 'sunk in the damp of a bog', cottagers with feus or long leases were careful to choose warm and dry places to build their houses: in the sheltered depths of a glen, beside a burn or up against a plantation of trees. Even when built on the edge of the muir for easy access to fuel, 'care is taken that they be sheltered and decorated, amid the blooming furze or the yellow broom'. The cottages themselves were old, low, often built of clay and turf, with grateless fires in both of the rooms, yet

> every cottager has a little garden [averaging one-sixteenth of an acre] … . And many of them bestow much care and show no little taste in its cultivation. Besides raising different kinds of coleworts [kale], cabbages, onions, carrots,

25 James Trotter, *GV West Lothian*, Edinburgh 1811, 140.
26 R. Douglas, *GV Selkirkshire*, Edinburgh 1813, 282.
27 R. Douglas, *GV Roxburgh*, Edinburgh 1813, 116.
28 Ibid., 116.
29 Headrick, *GV Angus*, 1813, 367.

> etc for the pot, they frequently have rows of gooseberry and currant bushes, together with roses and other flowering shrubs. Some of them decorate the walls of their houses with honeysuckles, or with ivy; and in some instances, with cherry and apple trees.[30]

They also grew herbs such as thyme, lavender, southernwood and tansy ('to kill the worms') and usually kept bees, as amusement, not for profit.[31] Robertson described these Mearns cottages all over again in his *Rural Recollections* of 1829. What particularly attracted him was the way the cottages clustered together but kept independent of each other 'and all front in different directions according as taste or want of taste dictates, each encircled by its own patch of ground'.[32] It was Robertson who noticed the neat little gardens of the little village of St Cyrus, alongside the highway to Aberdeen.[33] These cottages were on the Woodston estate, owned by William Adam of Kinross, who had insisted on dung heaps being moved to the back of the houses. 'The consequence is, that there is emulation for neatness'.[34]

Near Aberdeen some 'poor gardeners' sold their vegetables in the city – and paid very high rents for their ground.[35] It is not clear from this account whether or not they were market gardeners, who were quite common around Edinburgh and other large towns. But, north of Aberdeen, the picture changes. The Moray Firth coast was still quite poor and undeveloped, and the *GVs* do not mention any gardens.

JUDGEMENTS ON COTTAGE GARDENS

The highest accolade that could be bestowed on a village seems to have been to describe it as English. It was in this period that the civilising effect of growing flowers was being strongly promoted, and when England above all was seen as the home of flower-filled cottage gardens, unlike – according to various authorities – Wales, France, Portugal and the USA (the last two condemned by William Cobbett).[36] Flowers and flower-growing were believed to work on both the morals and the intellect of the poor.[37]

Several contributors to the *NSA* in the 1830s admired the Englishness:

---

30 Robertson, *GV Kincardineshire*, 185.

31 Ibid., 392.

32 Robertson, *Rural Recollections*, 416.

33 Robertson, *GV Kincardineshire*, 203.

34 William Adam, Letter to President of the Board of Agriculture on improvements on Blair, in David Ure, *GV Kinross*, 1797, Appendix, 8.

35 Skene Keith, *GV Aberdeenshire*, 157.

36 Keith Thomas, *Man and the Natural World: Changing Attitudes in England 1500–1800*, London 1983, 223ff.

37 Nicolette Scourse, *The Victorians and their Flowers*, London 1983, 53.

> [Kettins, Angus] is delightfully situated on a little stream, which, after passing through Cupar Angus, falls into the Isla. It is generally admired by strangers as a picture of rural simplicity and beauty. The neatly kept cottages and gardens – the village green, situated in the centre, watered by the rivulet, and the magnificent wood in which the whole is embosomed, give it much the appearance of the beautiful hamlets in the southern counties of England.[38]

Kingsbarns on the east coast of Fife was also reminiscent of an English village and 'much improvement in the condition and comforts of the people has of late years taken place. An inclination to render their persons more neat, their houses more cleanly, and their fore-grounds and gardens more ornamented, is prevalent'. The village was on the best land of the parish and had been greatly improved in the previous ten years by covering the drains and building metalled roads, and by the 'neat flower plots' in front of the new houses.[39] In the same vein, the minister of Lunan in Angus, as well as criticising the dire state of the cottages, bemoaned the lack of 'small and tasteful cottage gardens in front, which is often to be met with, and so universally admired in England', for which he firmly blamed the landowners and the farmers.[40]

The 1830s saw an increase in the approval ratings of cottage gardens. Perthshire cottagers seem to have had a particular penchant for gardening throughout the period. In Kincardine in Monteith there had been a horticultural association for some years[41] and there had been 'much improvement' in the gardens.[42] Kilgraston (also known as Kintulloch, Kuntilloch and now Kintillo), near Perth, was an unplanned village of old thatched cottages where 'the inhabitants, by training evergreen shrubs and China roses, and cultivating flowers before their houses, have given it a look of simplicity and cheerful gaiety that renders it positively picturesque.'[43] In a parish on the edge of the Highlands 'horticulture has of late made very considerable progress, and the several kinds of potherbs and roots, confined at a former period to the gardens of gentlemen, are now raised by all classes.'[44] Some of the cotton handloom weavers in the parish of Comrie 'enjoyed good gardens and a patch of potatoes and a pig'.[45]

The whole picture is hard to establish: were the pretty gardens rare gems in a sea of kale, boosted by writers intent on encouraging their occupiers, or was the taste for gardening spreading over the whole country? James Stuart Menteath's

38 *NSA* Kettins, Angus, vol. XI, 641.

39 *NSA* Kingsbarns, Fife, vol. III, 93–4, 90.

40 *NSA* Lunan, Angus, vol. XI, 331.

41 This was the Kilmadock and Kincardine in Monteith Cottage and Farm Garden Society and included the cotton mill workers of Deanston (see Chapter 7).

42 *NSA* Kincardine in Monteith, Perthshire, vol. X, 1280.

43 *NSA* Dunbarny, Perthshire, vol. X, 818–19.

44 *NSA* Monivaird and Strowan, Perthshire, vol. X, 732.

45 *NSA* Comrie, Perthshire, vol. X, 590.

1831 article on cottage garden societies opens with the lament that the gardens were 'generally neglected and ill-cultivated, and often overgrown with weeds and useless plants'.[46] The care taken in the *NSA*, in newspaper reports and the gardening press, to publicise villages or hamlets with well-tended and pleasing gardens raises the suspicion that they were still a small minority. Such articles tended to give landowners the credit. Under the heading 'Improvement in Scottish villages', a letter to *The Scotsman* in 1836 deploring the disreputable state of the mining village of Roslin in Midlothian earned a corrective editorial response describing four pretty and well-kept villages on different Lowland estates. Each landowner is named and their responsibility for the good appearance of the villages is strongly implied. Hermiston, just outside Edinburgh and owned by Sir James Gibson Craig of Riccarton, was 'a pattern of cleanliness and the houses are ornamented with shrubs and flowers'; new cottages in Lord Kinnaird's village of Inchture and at Craigend on the Moncreiffe estate in Perthshire had flower plots in front; Dirleton in East Lothian 'presents another fine specimen of cleanliness and rural ornament'.[47] In Kilgraston even the children were exceptional: there were no fences around the gardens 'yet they grow uninjured by children who were civilised and refined in their habits compared to other village children'.[48] In the *NSA*, the earl of Fife was praised for the 'neat cottages and lodges, with tasteful gardens' lining the turnpike road for five miles out of Banff in the north-east, otherwise unremarkable for pretty cottage gardens.[49]

John Claudius Loudon, the Scots-born horticultural author, botanist and landscape designer and ardent promoter of cottage gardening, focused more on the efforts of the cottage gardeners themselves. His *Gardener's Magazine* published frequent articles on how cottage gardens could be made more productive and more attractive, and also reprinted positive local press reports. In Fife 'it is no uncommon thing to see detached cottages, and cottage rows, embellished with flowering shrubs, and odoriferous, creeping and clinging plants.' Empty ground in towns and villages which had been 'really a nuisance' was now being planted with trees and shrubs. In new villages cottagers were planting decorative plants in 'a space ... lined off, in front of the houses ... and sorry we are when we find any of these desecrated to culinary purposes'.[50] The *Gardener's Magazine* also contained encouraging snippets about individual cottage gardeners of exceptional ability, such as the Elgin weaver whose grape vine, trained against his

46 Menteath, 'On cottagers' garden societies', 791.

47 Alexander Somerville considered 'sweet' Dirleton, a 'pet village', beautified to enhance the proprietor's mansion: out of sight of the mansion old, dilapidated and wretched cottages could be found amid luxurious farmhouses elsewhere on the estate (A. Somerville, *The Whistler at the Plough*, Manchester 1852, 173–4).

48 *The Scotsman*, 28th August 1836, 3.

49 *NSA* Gamrie, Banffshire, vol. XIII, 281.

50 'Domestic notices', *GM*, 1839, vol. XV, 195–8, 197.

house, had produced nearly forty clusters of ripened fruit.[51] Taste and skill could also now be found in some of the small gardens enhancing the appearance of new cottages along main roads on the earl of Aberdeen's estate in the north-east.[52]

Set against these pleasing vignettes are judgements like these: also in the *Gardener's Magazine*, 'An Englishman', once again contrasting the professional gardeners with 'the kail-yards of the peasantry', described the latter as 'the worst-managed, and the most tasteless in Europe';[53] a professional gardener, less intemperately: wrote 'in general, [cottage gardens in Scotland] are but indifferently kept' and overrun with weeds.[54] The *North British Agriculturist* (*NBA*) repeated the claim that although the Scots were 'a vegetable-feeding people' they grew only potatoes, onions, leeks and cabbage[55] – in general very badly, with not enough manure. 'A taste for flowers is in many districts unknown and consequently a Scottish cottage garden bears no comparison with' an English one. Some landowners tried hard to encourage better gardening but getting cottagers to accept their help was 'up hill work'.[56] William Cobbett had observed, as he travelled through East Lothian, 'you see no such thing as a little garden before the door; and none of those numerous ornaments and those conveniences about labourers' dwellings which are the pride of England', although he thought they cared for their children better than the English.[57]

THE WORKERS' EXPERIENCE

The writings of rural workers about their own gardens in the first half of the nineteenth century give intimate portrayals of the reality of cottage gardening as it was experienced day by day. Their reminiscences open the garden gate and sketch lively pictures of their plots and how they created them.

In the early decades of the nineteenth century the brothers Alexander and John Bethune, born into a desperately poor labouring family in north Fife, worked as labourers and gardeners and created new gardens out of weed- and rock-infested wastelands. Through their often unbearable struggles glows a passionate love for nature and for gardening: the never-ending backbreaking toil, the failed crops – but also the pleasure of creating one's own garden out of bare ground and begged

51 'Domestic notices – Scotland', *GM*, January 1828, 355.

52 James Black, 'Report on the cottage accommodation of the district of Buchan', *THASS*, 1853, 92–9, 96.

53 'Domestic notices', *GM*, 1832, vol. VIII, 470–6, 474.

54 Peter Mackenzie, 'Cottage gardening adapted to Scotland', *GM*, 1842, vol. XVIII, 315–20, 320.

55 The Englishman quoted above added: 'talk to a Scotchman about … [scarlet runners and vegetable marrows], and he will tell you the first is a poison and the second nothing but water', *GM*, 1832, 474.

56 *NBA*, 3rd January 1849, 13.

57 Cobbett, *Tour in Scotland*, 90.

Plate 12 Bethune Cottage, Newburgh. Author's photograph, 2021.

cuttings, success with new crops, the delight of spring growth. Alexander Bethune has provided one of the most appealing accounts of this exhausting yet joyful labour of garden creation. He and his younger brother John both published poetry and short stories in the 1830s, to some acclaim, while scratching a bare living, forced to take the most menial, poorly paid and dangerous work – labouring, weaving, stone-breaking and quarrying (where Alexander was badly injured twice in gunpowder accidents). Their efforts to raise themselves bore fruit when John was employed not long before his early death as head gardener on a neighbouring estate, helped by Alexander. Alexander's memoir of his brother recalls their early enthusiasm, labour and mistakes and provides an unforgettable picture of working-class gardening at this time.

On the day they left their old home near Loch of Lindores they dug up some of their favourite bushes and perennials to take with them to the house they had built on the hill above Newburgh:

> For a time, [John] stood with his spade in his hand, idly looking on … . 'I was just thinking … how many years, and how much toil and care it has taken to produce the scene before us; and now in a single half hour we will effectually mar its regularity and beauty; and before another year pass, perhaps, the last bush and the last flower may be rooted out by our successor, and the whole converted into a green-kail and potatoe field!'[58]

[58] A. Bethune, *Poems by the late John Bethune; with a sketch of the author's life*, 2nd edn, London 1841, 74.

In failing health, John then started work with Alexander on a garden for their new home. This involved cutting rock from the hillside and building a garden-dike. After John's death, Alexander tried to continue the work up to his own death at the age of thirty-nine. Bethune Cottage still stands, its sheltered garden behind it, high above the River Tay (Plate 12).

Alexander Somerville recalled how, in the winter of 1816–17, when he was five, the cows ate the family's small store of potatoes following a terrible harvest.[59] His family moved constantly around Berwickshire as his father sought work as a labourer. The family's garden was vital. One of their gardens was 500 square yards, 'our little portion of the land of Britain', and provided food all year round. He remembers such pleasures as birdsong, nosegays of flowers, and the occasional interloping (and doomed) hare, but 'we could not afford an inch of it to lie waste. We dug into the roots of the hedge, encroached on every thorn or holly-bush, above or below, until they were trained into respectable fences, seemly to look on, serviceable for their purpose, and unobtrusive.'[60]

The *Autobiography* of John Younger, the radical shoemaker, poet and angler of St Boswells in Roxburghshire, contains an account of equally heroic labour. At the age of only thirteen he set out 'to work out and enclose a small plot of waste front ground' at his family's new home. For this, he had to dig out stones and rubble to a depth of eighteen inches and then carry hundreds of bucketfuls of soil from behind the house. Next he laid out 'what I supposed a true garden style in miniature, with its box-edged gravel walks of a foot-breadth etc.' His efforts won him the approval of the irascible local laird. In spite of his care for design, however, the only crops he mentions were his very successful cabbages.[61]

INDUSTRIAL WORKERS' GARDENS

The conditions in which handloom weavers worked, in loom sheds tacked onto their cottages, gave them a different view of the garden from that of the labourers, although they alternated their spells at the loom, mostly in winter, with harvest and other farm work. They had time for the garden and some also rented additional land.

> I may very safely say that in Angus shire only there are employed in the linen manufacture from 20,000 to 25,000 people. Many of these reside in the towns of Dundee, Arbroath, Brechin and Forfar but many, and I believe the greater number, are scattered through the villages and the country where every one has a garden and generally an acre or two of ground.[62]

---

59 A. Somerville, *Autobiography*, 18.

60 A. Somerville, *Whistler at the Plough*, 210.

61 John Younger, *Autobiography*, Kelso 1881, 106.

62 Headrick, *GV Angus*, 1813, 126.

Handloom weavers in both town and country at this time were lauded as prolific readers, writers and self-educators, and often feared as political agitators and debaters. They were also keen gardeners. In towns and cities the most skilled and best-paid formed clubs dedicated to raising new varieties of flowers. In the countryside, less skilled and very rarely full-time, weavers' gardens supplemented their income, as a slightly fanciful newspaper report explained:

> The weavers [of Cupar] who are lucky enough to have a garden with their cottages, are now [in March] full of business and thrift, labouring morning and evening to commit their seeds to the ground, and looking forward to rich crops of onions, carrots and early potatoes; they expect renovated health as well as easier subsistence, from their industry; and pity their brethren in towns who cannot have the treat of once a day exchanging the narrow and gloomy workshop for the companionship of flowers and blackbirds in their little gardens.[63]

The blessings of fresh air and exercise is a common theme in descriptions of weavers' gardens. Here, they could relieve the tedium and the damp and dusty atmosphere of their loom sheds, and the isolation. We often read of weavers gardening together, as in Cupar. Half a dozen wool-combers too were discovered in Aberdeenshire 'at work by four o'clock in the morning, hoeing potatoes for family use' and for their health.[64]

Many of the weavers of Arntully (or Airntully), Perthshire, were also 'pafflers', renting smallholdings. In fact the minister who wrote the *OSA* account was rather at a loss as to whether they were tradesmen or farmers. Each had 'a small bit of ground' attached to their houses which they used to grow 'garden stuffs' and grass for the cow.[65]

The arrival of factory weaving in the 1830s brought desperate poverty to handloom weavers as their wages collapsed. In the countryside this was relieved somewhat by their gardens and potato grounds. Rural linen handloom weavers, including women and teenagers, who wove coarser cloth, had some reprieve as this cloth demanded less skill and could not at first be woven by the power-looms. The Airntully weavers still held on to their gardens, according to the *NSA*, and to their cows, but the village had shrunk and was expected to disappear.[66] Decline was inexorable, and by 1900 there were no handloom weavers left in the countryside.

---

63 *Fife Herald*, 24th March 1831, 3.

64 Skene Keith, *GV Aberdeenshire*, 157.

65 *OSA* Arntully, Perthshire, vol. XIX, 329, 332.

66 *NSA* Arntully, Perthshire, vol. X, 1135. It was still there, however, and still apparently inhabited by linen weavers in the 1880s (Arntully, in *Groome's Ordnance Gazetteer of Scotland* 1882, available at http://gazetteerofscotland.org.uk.)

One of the legacies of the rural handloom weavers in the east of Scotland is the poetry of the Bethunes, John Nevay, James Thomson of Kenleith and many others, which often testify to the power of gardens to charm, inspire and amuse them. Alexander Laing (1787–1857), from Brechin, Angus, warned the moles in his grandmother's garden, or kailyard (which fits the rhyme better and adds a nostalgic note):

> And whan John Frost agen grips hard
> Ye may tak the beild o our kailyard – [shelter]
> Till kindly spring thaw bank an' brake;
> Till ye hear the dinnle o' spade an' rake. [vibration]
> Our cabbage and greens maun grow for you,
> Our syboes an' leeks that season our broo,
> Till simmer is ower maun a' be spared,
> If ye wad *Winter* in our kailyard.
>
> The Mowdiwort[67]

Rural water-powered mills producing cotton, linen and woollen yarn and paper sprang up from the 1780s across the Lowlands but those about which most is known are Stanley and Deanston, both cotton-spinning mills. Today the mill buildings and the villages built to house the workers still stand, although the mills have fallen silent. All the village houses had gardens, although mill workers can have had little time for gardening: they worked, in extreme heat, for twelve or thirteen hours a day, and only slightly fewer on Saturdays. Hours came down slowly through the century: by the end of the century fifty to sixty hours a week was the norm.[68]

Stanley Mills opened in 1784 on the banks of the River Tay in Perthshire. The workers' village was built beside the mills on poor agricultural land leased from the duke of Atholl. The mills had uneven fortunes but in the 1830s 1,000 workers, many of them children, were employed, in relatively wholesome conditions, according to the *1st Report* of the 1833 Factory Commission. There were WCs on every floor of the mill, and pot plants on the windowsills.[69]

The original partnership included Richard Arkwright, inventor of the spinning frame and the 'father of the modern industrial factory system'; George Dempster of Dunnichen, the MP and arch-improver with a finger in every progressive pie; and the duke of Atholl, one of the largest landowners in Scotland, also famed for improving his estates. Arkwright left the partnership before production began, but not before leaving his mark on the mill, modelled on his Cromford establishment

67 Alexander Laing, *Wayside Flowers*, 4th edn, Edinburgh 1878, 57.
68 Smout, *A Century of the Scottish People 1830–1950*, London 1986, 102.
69 *Factories Inquiry Commission, 1st Report*, 1833, 62.

in Derbyshire.[70] At Cromford, Arkwright had stipulated that gardens should be attached to the workers' houses.[71] While Arkwright's involvement in the layout of the village at Stanley is not known, all the houses did have attached gardens. The duke had learned from the example of planned villages such as Laurencekirk and designed a grid of straight streets and squares which were to be lined with one- and two-storey houses 'in one continuous line without any empty spaces intervening, excepting entries of four foot or thereby where necessary to the gardens behind'.[72] Whether the inhabitants were expected to grow vegetables to supplement their wages is not known, nor do we have any more information on the gardens. There does not seem to have been a horticultural association here until much later and the *NSA* has little to say about the mill workers beyond classifying them as 'not a reading community' and having too many alehouses.[73]

Deanston Mills on the River Teith near Stirling were also much admired by the Factory Commissioners in 1833. 'This is one of those magnificently situated and admirably regulated great manufacturing establishments, which it is a pleasure to see.' The workers also impressed the Commissioner: 'a more cheerful, happy-looking set of industrious men and women, and of young people, is seldom, if I am not mistaken, to be found.'[74] The workers were housed in what eventually became three 'divisions' or blocks of three-storey terraced tenements alongside the river, still standing and occupied today. The *NSA*, quoting from the *Inverness Courier*, described the village as it was in the 1830s in these approving words:

> Immediately adjoining the works is a handsome little village, built and founded by the company, which contains about 1200 inhabitants. The houses are neat, built in one long street parallel to the water course, and are two stories high, with attics. They are most exemplary patterns of cleanliness, and to each house is attached a small piece of garden ground, and a range [sic] of grass plot for bleaching.[75]

---

70 A. J. Cooke, *Stanley – From Arkwright Village to Commuter Suburb: 1784–2003*, Perth 2003, 58.

71 Willes, *Gardens of the British Working Class*, 134.

72 Cooke, 'The early development of Stanley 1784–1823', in A. J. Cooke (ed.), *Stanley – Its History and Development*, Dundee 1977, 14. The 1st edition Ordnance Survey map shows great variety in size: nine houses with front and back gardens in one street covered nearly three and a half acres. A more densely populated square of approximately fifty houses was just over two acres. Ordnance Survey 1st edition 25 inches to the mile, Perth and Clackmannanshire CXXV.13, surveyed 1862.

73 *NSA* Auchtergaven, Perthshire, vol. X, 447, 449.

74 *Factories Inquiry Commission*, 16.

75 *NSA* Kilmadock, Perthshire, vol. X, 1233.

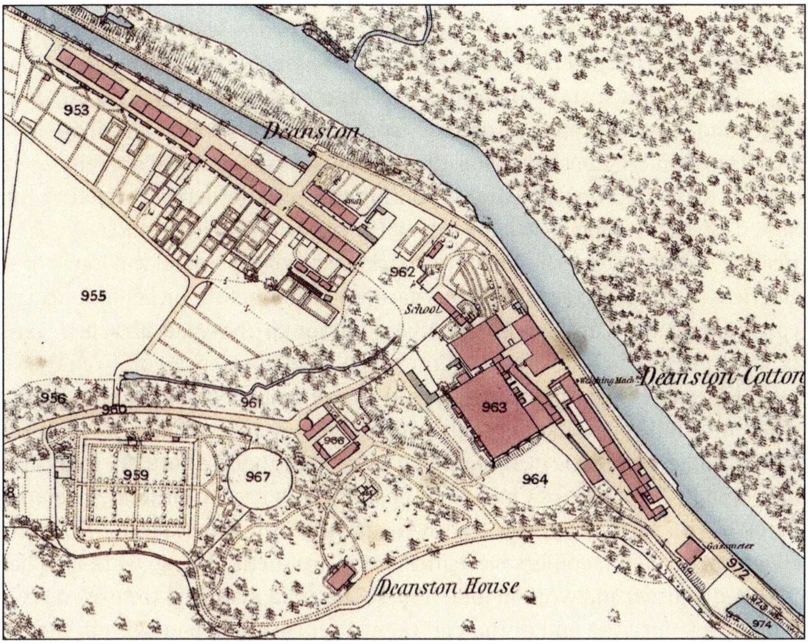

Plate 13 Deanston. Detail of Ordnance Survey, 1st edition, 25 inches to the mile, Perth and Clackmannanshire CXXV.13, surveyed 1862. Workers' tenements and gardens are at the top left.

There were also small gardens opposite the tenements on the river bank, later turned into a bowling green (Plate 13).[76] More surprisingly, on the roof of the weaving shed soil to the depth of three or four feet had been laid as insulation and a flower garden had been planted[77] – although when John Claudius Loudon visited in 1841 it was 'covered with grass, and either was, or is intended to be, grazed by sheep'.[78] We know that the Deanston cotton mill workers gardened because they started, with the support of the mill owner, the inventor James Smith, one of the earliest cottage garden societies in the 1830s.

Until 1799 most Scottish miners, especially in the east, were bound to their masters as serfs,[79] living in appalling conditions and shunned by the rest of the countryside. Emancipation brought little improvement. The shocking details of

76 Personal communication from Kenny Lauder, 23rd October 2019.

77 *NSA* Kilmadock, Perthshire, vol. X, 1237.

78 J. C. Loudon, 'Manchester, Chester, Liverpool and Scotland in the Summer of 1831' in *Gardening Tours 1831–1842*, available at https://www.gardenvisit.com/book/gardening__tours_by_jc_loudon_1831-42.

79 Coal owners had complete freedom to recapture runaway colliers, even after many years, to move them from one pit to another and to bind their newborn children to work in the collieries (Smout, *Scottish People*, 430–3).

the barely imaginable misery and filth of the colliery rows were recorded in Edwin Chadwick's *Sanitary Report* of 1842 and the 1844 *Report on the State of the Population in Mining Districts*. Amidst such conditions, matched only by their working conditions underground, gardens and gardening would seem a luxury, although before emancipation miners' houses in Alloa, Clackmannanshire – and probably elsewhere – had had attached gardens.[80] Loudon, on his tour of southern Scotland in 1841, remarked of the mining district west of Dollar, Stirlingshire, which was also a flourishing farming area: 'the only drawback we recollect was in the long rows of comfortless-looking cottages, without gardens either before or behind, occupied by colliers.'[81] In the Lothians in the 1840s, although they had attached garden plots, 'it is very rare to find any one of the respective plots fenced off from the rest. The whole space is left open and exposed to the incursions of children, poultry, or other intruders. Little care is consequently bestowed on what is so liable to be injured.'[82] Colliers' gardens in Tranent and Musselburgh, East Lothian, at this time, as inspected by Dr Alison for Chadwick's *Sanitary Report*, were either neglected and overgrown with weeds or covered with 'heaps of putrid materials'.[83]

Like farm servants, colliers were inclined to frequent flitting, were also liable to instant dismissal and worked long hours. They had managed to resist a six-day working week until the end of the eighteenth century, and had been in the habit of taking frequent holidays.[84] After emancipation hours lengthened. Miners were underground for sixty hours a week in 1850. Later, the Fife miners won an eight-hour day in 1871 and the average for the east of Scotland by the 1890s was under forty-four hours per week.[85]

Precious free time spent on a garden whose fruits they might never enjoy would have been deemed wasted. The 1844 Report's author, the Commissioner H. S. Tremenheere, points out, however, that English miners suffered from short hires too, 'yet … every plot of garden-ground … is scrupulously fenced round, and every inch of the soil diligently cultivated'. He did find one fenced garden-plot, only to be told that it belonged to an Englishman. His solutions were to bring in a regulation requiring the inhabitants to keep their poultry from straying and to encourage colliers' wives and children, now freed from underground work thanks to the 1842 Mines Act, to cultivate allotments as a way of making up for

80 *OSA* Alloa, Clackmannan, vol. VIII, 615.

81 Loudon, *'Manchester, Chester, Liverpool and Scotland in the Summer of 1831'* in *Gardening Tours 1831–1842*, available at https://www.gardenvisit.com/book/gardening__tours_by_jc_loudon_1831-42.

82 *Report of the Commissioner Appointed under the Provisions of the Act 5 & 6 Vict. c.99, to inquire into the operation of that Act and into the state of the population in the mining districts*, by H. S. Tremenheere, 1844, 13.

83 Edwin Chadwick, *Report on the Sanitary Condition of the Labouring Population of Great Britain* [1842], ed. M. W. Flinn, Edinburgh 1965, 87, 135.

84 Robert Duncan, *The Mineworkers*, Edinburgh 2005, 46.

85 *RC Labour, Digest of Evidence, Vol. III, Mining, Iron, Engineering and Hardware*, 1893, 13.

the loss of their wages.[86] It is striking how often contemporary observers worked out ingenious solutions to workers' problems that depended entirely on extra effort by the workers themselves.

By the late 1860s Midlothian miners, if they gardened, were raising cabbages and leeks. Flowers were rare, and more often found in a 'dilapidated tea-pot on the window-sill' except outside the newer dwellings. At the marquess of Lothian's Newbattle colliery, however, where both the worst and the best housing could be found, there were flower gardens in front and kitchen gardens behind.[87] A few years later Scottish miners' houses were declared by an English MP to be 'no better than pig-styes'. This spurred the *Glasgow Herald* to 'instruct one of our staff to make a run through several of the mining villages, and procure at once such information as the miners themselves or their employers are able to furnish'.[88] The intrepid reporter's journey of exploration into the little-known territory of the coalfields and mining villages produced a series of detailed, lively and critical dispatches that are full of interest.[89]

The *Glasgow Herald*'s reporter managed to visit all the main coalfields in a few weeks, in February 1875. In the Lothians, then a small coalfield employing only 2,000 men, he found a peaceful set of workers, disinclined to strike and difficult to unionise, living in good-quality and clean houses (much better than in the west) decorated with pot plants and caged birds. Many had flowers growing outside. In one house he even noticed 'a small conservatory at the extremity of the kitchen garden behind'.[90] Nearly all the houses had kitchen gardens in front or behind, and these seemed generally to be cultivated, if not always very well.

Donibristle in Fife had very untidy back gardens, but the gardeners of the tiny nearby village of Fordell, now disappeared, had an enviable reputation. Tom Hunter recalls that 'it was a renowned Sunday summer walk for people in the neighbourhood to look at the Fordell roses'.[91] Our reporter was unimpressed:

> Fordell was represented to me as being a model village, belonging to Mr Henderson, of the Fordell Company. Simple candour, however, compels me to say that it is nothing of the kind. It is true that there is one very fine row of houses which we reach through long kitchen gardens and neatly arranged flower-plots. These are rooms and kitchens, entered from opposite sides of a passage, which is ventilated by doors in front and rear. They are large and well-finished houses,

86 Ibid., 14.

87 David Bremner, *The Industries of Scotland – Their Rise, Progress and Present Condition*, [1869], Exeter 1969, 27–8.

88 *Glasgow Herald*, 11th January 1875, 4.

89 Anon. [Glasgow Herald reporter], 'Notes on Miners' Houses', 1875. Available at the Scottish Mining website, http://www.scottishmining.co.uk/399.html.

90 Ibid., pt IX.

91 T. F. Hunter, *Mining in West Fife: the Crossgates Area in particular Fordell and Halbeath Collieries*, Norwich 2001, 257.

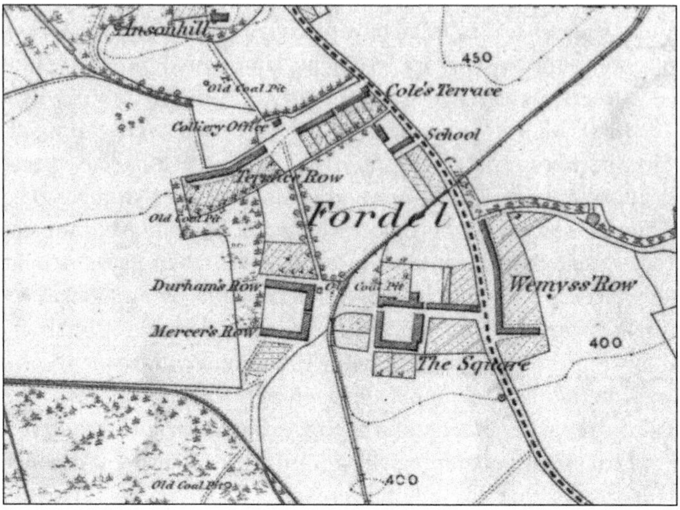

Plate 14 Fordell. Detail of Ordnance Survey, 1st edition, 6 inches to the mile, Fife, sheet 35, surveyed 1853.

> the rent being 6s 8d a month. Mr Henderson, I am told, stimulates a love for flowers by offering prizes in summer for the best kept gardens, and I have no doubt the result of the competition thus created is abundance of colour and fragrance; but as there are no outhouses of any kind, the chances are that the odours in warm weather will be unpleasantly mingled.[92]

The houses he went into were small, old and damp, although their front gardens were relatively large. They had measured 600–800 square yards (eighteen to twenty falls) in the 1840s, and were larger by the next decade (Plate 14).[93]

We have more information about gardening in Fordell in the late nineteenth century than for any other mining village in the east. Bob Holman's touching reminiscence of the place and the people in the 1890s, perhaps somewhat gilded by the passage of time, relates that Fordell, contrary to the *Glasgow Herald*'s view, 'has always been known for its floral beauty'.[94] The Henderson family, mining the coal on their land since the late seventeenth century, introduced a best-kept garden competition in 1859, for which nearly every garden in the village was entered, and a highly popular flower show. Young married couples received a

92 'Notes on Miners' Houses', http://www.scottishmining.co.uk/399.html.

93 Hunter, *Mining in West Fife*, 253.

94 Bob Holman, *Behind the Diamond Panes: the story of a Fife mining community*, Cowdenbeath 1952, 42.

cartload of manure from Fordell Castle as a wedding present and 'the "gairden" took the foremost place in the life of the village.'[95]

In Midlothian, paternalism had a rather different face: no wedding presents of dung, but greatly improved housing. Both the marquess of Lothian and Archibald Hood built new miners' rows from the 1840s, provided piped water to wells, drained the streets and later supplied gas. The rows Hood built in Rosewell in the late 1850s and after had large back gardens. In the marquess of Lothian's Newtongrange, the newest houses, built from the 1870s and now of four to five rooms, had piped water, tiled floors and slate roofs and flower gardens in front with 'kitchen gardens' around fifty-five feet long behind. Both the houses and the gardens were regularly inspected and, if found to be lacking in cleanliness, the occupier could be sacked.[96] This harsh rule suggests one purpose of the miner's garden: to reinforce workforce discipline. Archibald Hood, notorious as an authoritarian owner, was not showing a softer side in promoting gardening and founding gardening clubs and flower shows. Just like the Hendersons, he regarded a well-tended garden as an outward and visible sign of an obedient and compliant employee.

By the early twentieth century Newtongrange had become a small town, owned almost wholly by the partnership of Lothian and Hood. Rules for punishing lazy gardeners were now slightly modified: the colliery gardeners dealt with untidy gardens and the work was paid for out of his wages by the defaulting miner.[97] But by the time of the 1912 government inquiry into working-class housing in Scotland (not published until 1918) the overall picture of miners' rows was still squalid:

> The 'Miners' Row' of inferior class is often a dreary and featureless place, with houses, dismal in themselves, arranged in monotonous lines or in squares. The open spaces are encumbered with washhouses, privies, etc., often out of repair, and in wet weather get churned up into a morass of semi-liquid mud, with little in the way of solidly constructed road or footpath – a fact which adds greatly to the burdens of the overwrought housewife.[98]

Nevertheless, gardening, the same report approved, is 'in some mining communities… highly developed'.[99] This depended on the family's having a long enough tenure of their house and garden, the proximity of the garden to the house, and adequate fencing. Where the miners owned their houses, as in Kelty in Fife, which had a successful flower show, or where – as in East Wemyss and Buckhaven also in Fife – several generations had occupied the same house, or an improved house on the

95 Ibid., 68.

96 S. Martin Gaskell, *Colliery Housing: a study in Midlothian*, Edinburgh 2011, 8.

97 Ibid., 19.

98 *RC Housing of the Industrial Population of Scotland*, 1917–18, 5.

99 Ibid., 127.

same site, good gardens could be expected.[100] In most of Fife, however, there was a lack of provision of garden ground and miners had little taste, or energy, for gardening. As with the farm servants, those observers who chose to examine the miners' situation closely were quite aware that good gardens came from security, comfort and free time – not from exhortation and bullying.

PAUPERS AND LIGHTHOUSE KEEPERS

Two groups at opposite ends of the Victorian working class had gardens provided to ensure subsistence: sick and elderly paupers in parish poorhouses and lighthouse keepers in onshore lighthouses.

Before the Scottish Poor Law was reformed in 1845 arrangements for poor relief were the responsibility of the Church elders, who collected voluntary contributions and fines from parishioners and doled out allowances to those in need, mainly the elderly, the sick and children. In Scotland the able-bodied poor had no entitlement to relief. A few parishes, thanks to generous endowments, provided individual or communal houses for paupers rent-free, with attached gardens, similar to English almshouses. At Largo in Fife, twelve old men all called Wood lived in the hospital founded by John Wood and ate the vegetables grown in the large garden.[101] In Whitekirk, East Lothian, the paupers' houses and gardens in the Widow's Row, gifted by the earl of Haddington, were much sought after 'even on the part of healthy, young, and vigorous persons'.[102] A handful of other parishes offered similar support.

Under the new Poor Law the unemployed were still denied help but good care was often taken of the old, infirm, widows and children, increasingly financed by compulsory assessment. Hospitals and dispensaries were opened and poorhouses built for the elderly and sick.[103] In Aberdeenshire these often had gardens. A landowner in Culsalmond gave one eighth of an acre as a garden for the parish poorhouse in 1857.[104] The Lonmay poorhouse, built at the end of a track in a remote corner of the parish, had a 'considerable' garden.[105] No evidence has come to light revealing how and by whom the gardens were tended. Unless the parochial

[100] The *Glasgow Herald* reports state that cheap rents were 'ridiculously low' in many villages in the Lothians in the 1870s. But, in spite of impressive savings, the miners had little interest in buying their homes – except in Tranent, East Lothian, where there was a large and prospering co-op ('Notes on Miners' Houses', http://www.scottishmining.co.uk/399.html).

[101] *OSA* Largo, Fife, vol. IV, 543.

[102] *NSA* Whitekirk and Tyninghame, East Lothian, vol. II, 41.

[103] Rosalind Mitchison, *Life in Scotland*, London 1978, 131–3.

[104] Aberdeen City and Aberdeenshire Archives, AS/Apcul/1, Culsalmond Parochial Board/Parish Council Minute Book, 5th January 1857.

[105] Aberdeen City and Aberdeenshire Archives, AS/Aplon/1, Lonmay Parochial Board/Parish Council Minute Book, 23rd May 1848.

board employed a gardener it is hard to imagine who was fit enough to garden and how well they were able to do it, but the provision of gardens does demonstrate some concern for the health and well-being of the inmates.

Lighthouse keepers formed a tiny, carefully selected and well-paid elite. They were usually recruited from amongst seafarers and skilled artisans such as smiths and joiners. As such, they are quite untypical of the mass of cottagers we are looking at in this chapter. However, keepers were bound to their lighthouses by their remoteness and the demands of their responsibilities and their gardens were as vital as those of farm servants and cottagers, if not more so. Lighthouse keepers were subject to strict discipline and close scrutiny by the Northern Lighthouse Board (NLB). Their gardens and gardening practice were highly regulated, and the NLB reports provide invaluable documentary evidence on both of these.

Each onshore lighthouse was manned by a principal keeper and one or two assistants. The principal keeper was paid, in the 1850s, £55 per year; assistant keepers had £46 per year.[106] They also received pensions on retirement. Free housing, fuel, a garden and some arable and pasture were added to this. In return, they were expected to be 'sober, industrious, cleanly in their persons and linens and orderly in their families' and were also required to be civil to tourists visiting the lighthouse.[107] Rules were exacting, and failure to keep the light burning punishable with instant dismissal.

There were thirteen coastal lighthouses on the east coast between the Moray Firth (Covesea Skerries) and the English border (St Abbs Head), most of them built by the Stevenson family. Most were sited far from towns and villages and surrounded by gardens, grazing ground and sometimes fields for cereal crops, sheltered by high stone walls. The NLB, usually in the person of a Stevenson who undertook annual trips of inspection, was at great pains to ensure that the garden ground was fit for purpose. When the lighthouse at Covesea Skerries was being built in 1846 the ground was, reported Alan Stevenson, 'of very poor quality consisting chiefly of sand and boulders with a slender sprinkling of light grass and before Gardens of any use can be formed, soil must be transported to the extent of about 1,600 cubic yards'.[108] The cost of preparing this ground for the garden was estimated at a staggering £150.[109] The site was lashed by high winds and so extra high walls were built.

At Girdleness Lighthouse in Kincardineshire the garden and 'park' for grazing, for two keepers, extended to ten acres.[110] A quarter of an acre of garden per keeper seems to have been the usual allocation. The NLB's *Instructions to*

---

[106] The average farm servant's wage was something over £24 at this time.

[107] Business papers of the Stevenson family, NLS, Acc 10706/525/25 1843 Instructions to light-keepers, 6.

[108] NLS, Acc 10706/95/59.

[109] NLS, Acc 10706/303, Annual Reports on Northern Lighthouses.

[110] *NSA* Nigg, Kincardineshire, vol. XI, 202.

Light-keepers, a small manual originally devised by Robert Stevenson, updated regularly and covering every aspect of a keeper's work and daily life, makes clear the importance the NLB attached to the garden ground and the arable land and pasture – not just for feeding the families but also as an inducement for them to stay in often inhospitable places. A keeper had to maintain the 'arable portion in a proper state of cultivation, by manuring, ploughing, and rotation of cropping, and he should not sow the same kind of crop on the same ground for two years in succession'.[111] Each year the NLB sent out the Oliver and Boyd *Almanac* to the lighthouse libraries and recommended the 'Farmer's Kalendar' or the 'Gardener's Kalendar' in it for those who might have no gardening experience. The NLB also supplied seeds for vegetables. Precise directions were given in the booklet as to the fate of crops when a keeper's two-year stint came to an end. If the crops had been sown but not yet harvested the incoming keeper was to compensate his outgoing colleague for seed, manure and labour. If the produce was ready for picking, the incoming keeper was obliged to buy it at a valuation.[112]

In 1852 a change of keeper at Covesea Skerries revealed an 'irregular agreement' by which some of the lighthouse ground had been let to a neighbouring farmer, perhaps taking advantage of the new topsoil. On this occasion, the principal keeper pleaded ignorance of the ban on subletting and got off with a severe reprimand, but subsequent editions of the *Instructions* spelled out a new rule: ground could not be sublet, nor could crops be sold, except to an incoming keeper. The 1867 edition also put the principal keeper in charge of dividing the available land amongst the keepers and regulating the numbers of sheep and cattle each man could pasture.[113]

Lighthouse keepers do not seem to have been natural gardeners. There is a note of exasperation in the letter the NLB sent to principal keepers in 1896, scolding them for exhausting the garden soil by overcropping with potatoes and reminding them of their duty to keep it well manured (Plate 15). On the other hand, the importance the NLB itself attached to the keepers' land is underlined by the regulation relating to lighthouses where there was no available ground. In that case, each keeper would receive a grant of £10 in lieu of land and £3 in lieu of the garden. 'This allowance is made to enable the light-keepers to purchase, as far as possible, milk, meal, potatoes and fresh vegetables, such as the possession of a portion of land would have yielded to them and their families.'[114] In the 1840s the potato blight brought near famine to parts of Scotland and Alan Stevenson

[111] NRS, NLC 5/5/5, Northern Lighthouse Board, Instructions to light-keepers, 3rd edn, Edinburgh 1874, chapter XVIII, 69.

[112] Ibid., chapter XVI, 3.

[113] Ibid., chapter XVI, 1.

[114] Ibid., chapter XVI, 6.

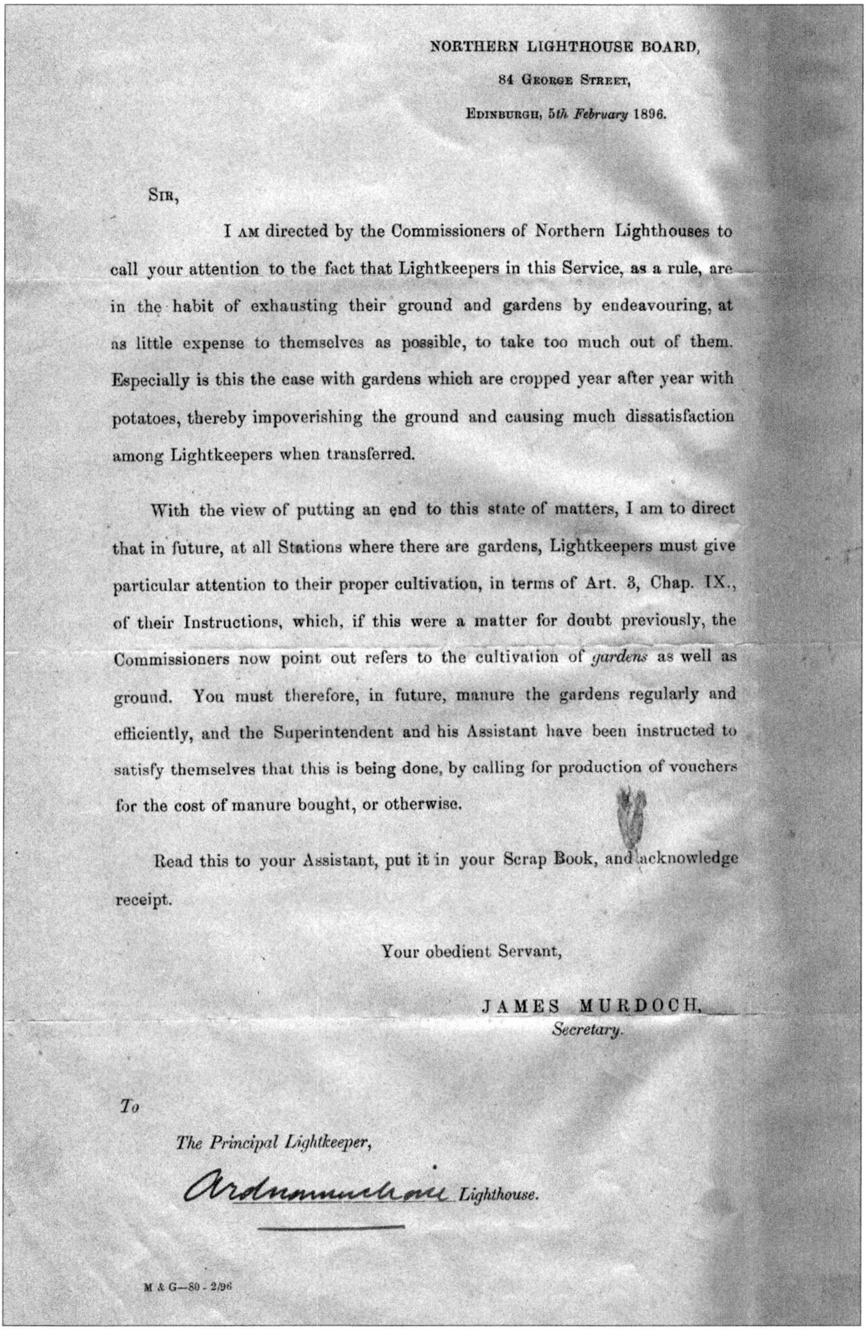

NORTHERN LIGHTHOUSE BOARD,

84 George Street,

Edinburgh, *5th February* 1896.

Sir,

I am directed by the Commissioners of Northern Lighthouses to call your attention to the fact that Lightkeepers in this Service, as a rule, are in the habit of exhausting their ground and gardens by endeavouring, at as little expense to themselves as possible, to take too much out of them. Especially is this the case with gardens which are cropped year after year with potatoes, thereby impoverishing the ground and causing much dissatisfaction among Lightkeepers when transferred.

With the view of putting an end to this state of matters, I am to direct that in future, at all Stations where there are gardens, Lightkeepers must give particular attention to their proper cultivation, in terms of Art. 3, Chap. IX., of their Instructions, which, if this were a matter for doubt previously, the Commissioners now point out refers to the cultivation of *gardens* as well as ground. You must therefore, in future, manure the gardens regularly and efficiently, and the Superintendent and his Assistant have been instructed to satisfy themselves that this is being done, by calling for production of vouchers for the cost of manure bought, or otherwise.

Read this to your Assistant, put it in your Scrap Book, and acknowledge receipt.

Your obedient Servant,

JAMES MURDOCH,
Secretary.

To

The Principal Lightkeeper,

Ardnamurchan *Lighthouse.*

M & G—80 - 2/96

Plate 15 Letter from the Secretary of the Northern Lighthouse Board to Principal Keepers, 5th February 1896.

Plate 16 Kinnaird Head lighthouse garden, Fraserburgh, Aberdeenshire, no date.

Plate 17 Cottages, Wormit, Fife, by Alexander Wilson, active in the late nineteenth century.

called on the NLB to pay each keeper £5 to £10 in compensation, depending on how far the lighthouse was from the nearest market.[115]

Apart from potatoes we have no records of what the keepers grew. The lengthy list of stores that they could order from the NLB, printed at the back of the *Instructions* manual, includes only a 'set of garden seeds'. The photograph of the garden at Kinnaird Head (Plate 16) appears to show several types of vegetables, but no flowers.

Today, all that remain of the old gardens of the now automated lighthouses are the high and sturdy walls, built to shelter the crops from wind and waves. The attention paid by the NLB in the nineteenth century to lighthouse gardens reveals not only a somewhat obsessive attitude to the welfare and behaviour of the keepers but also how vital the produce of the garden was for the diet and health of families stuck on their headlands, far from shops and markets.

GARDENS FLOURISH

How did the more prosperous and settled decades of the second half of the nineteenth century affect the cottage gardeners and their gardening? Not all rural workers benefited from rising living standards, but many did. There may not have been a direct relationship between better housing and gardening practice, as contemporary observers insisted, but it seems very likely that more comfortable houses, together with rising living standards, shorter working hours for many, greater security of tenure and the effect of these on workers' morale would have encouraged them to take more care of the home, to furnish and decorate it with pictures, china and pot plants inside and to care more for the garden outside (Plate 17). With the expansion of the railway network, the penny post and cheap newspapers came the circulation of gardening knowledge and opportunities to visit nurseries, parks and gardens and flower shows. Gardening for pleasure became both possible and popular.

There are few general descriptions of the cottage gardens after the 1850s, apart from comments in reports of village flower shows. Also missing, as Scotland became less of an alien territory to outsiders, are travellers' journals. It is possible that this in some way was related to greater social stability and the lessening of fear of the poor amongst the middle and upper classes. The political threat felt to have been posed by labourers and handloom weavers during the turbulent years of rural riots and Chartism in the 1830s and 1840s had gone. The housing of the poor remained an anxiety, however, prompting many Royal Commissions and government reports as well as philanthropic campaigns. All these inevitably touched on workers' gardens.

[115] NLS, Acc 10706/95/334 8/1/1847, Stevenson business papers, Annual Reports on Northern Lighthouses.

INSIDE THE COTTAGE GARDEN

The position and layout of the cottage garden can tell much about its function. In the eighteenth century the garden was at the rear of the house because the midden was at the front, conveniently just outside the only door. Early nineteenth-century cottages often had the garden in front, as recommended by Sinclair and Neill, so that the midden could be moved to the back, and out of sight. Later, it became usual to have a small flower garden in front and a vegetable garden at the back, along with the 'offices' – a privy and the ashpit that was slowly replacing the midden. Pigsties, hen houses and occasionally bee hives were also at the back. Bee-keeping was strongly encouraged by the Highland and Agricultural Society of Scotland (HASS) as a source of extra income, although it never seems to have been particularly popular. Later in the century pigeons and rabbits might be added to the garden menagerie. The whole was enclosed by a fence or stone dyke, and trees planted by the fence.

Tiny clues from the early decades of the period suggest that, where they tried to design their gardens, cottage gardeners were probably following a fashion of much earlier times – the early eighteenth-century style despised by Patrick Neill as 'stiff and formal' – laid out in squares with 'herbaceous plants disposed in the most precise arrangement'.[116] In St Cyrus 'every cottager has his garden laid out in little walks, rose-trees and flowers on the border, and his pot-herbs in the squares within',[117] while John Younger's juvenile attempt at 'true garden style' also produced little box-edged paths.[118]

We know in detail a handful of gardens displaying quite different attitudes to gardening and to their purpose. None may be said to be representative, but the descriptions convey strong impressions of what they looked like. The first garden Alexander (1804–1843) and John (1812–1839) Bethune created at their family home was no ordinary cottage garden, although it also featured box edging; but, before they tackled it, it resembled nothing so much as a kailyard:

> The garden, or yard, as it was called, which belonged to the house when we came to it [as children] was one of the most worthless of the kind. The soil, for the most part, consisted of only a few inches of earth, with a substratum of gravel and large stones; it had no fence except the ruins of what had once been a sort of dike, from which a luxuriant crop of nettles was annually produced, and there was neither bush nor flower within its boundary. Almost from boyhood we had been eager to bring it into better condition; but, like those who must acquire knowledge by experience without the aid of instruction, at first we were sad bunglers, and our progress in the way of improvement was necessarily slow ... The ruins of the old dike had been

116 Neill, *Gardens and Orchards*, 3.

117 Adam, 'Improvements on Blair', 8.

118 Younger, *Autobiography*, 106.

> razed out, the nettle-roots extirpated, and a new stone fence built in its stead; a very considerable portion of the ground had been trenched over, and many cart loads of stones taken out before a sufficient depth of soil could be obtained; a bee-house, as it was called, had been erected, with a small semi-circular green in front of it, surrounded a gravel walk with box edgings. Beyond this was a narrow border, bounded also by a curve, which served as a flower garden; and, to enclose and shelter the whole, a double row of currant bushes – one on the out and another on the inside – had been trained to a rustic railing. Within the same enclosure, a framing of an arbour had been erected, and honeysuckle and ivy planted around it ... Two or three years before we left the place, well trained and regular-looking rows of gooseberry bushes on either side of the walks, or rather alleys, had come into full bearing. The ground had become fertile in proportion to the labour and manure bestowed upon it. Almost every season had added something to our collection of flowers. Before the close of the last summer of our sojourn there, the longest shoots of honeysuckle had met upon the top of the arbour, giving promise of an umbrageous roof ...[119]

The dedication, determination and huge physical effort required to clear the ground and build the walls was matched by the brothers' aesthetic sense that designed the curving flower bed and placed the creeper-covered arbour.

Nelly Johnstone, an old woman who made a living from spinning and as a herbalist, had a garden in the village of Longforgan in the Carse of Gowrie, Perthshire, around the middle of the century. Defying both the elements and a lack of proper tools, hers was a more typical cottage garden. She lived in a one-room house made of clay-and-bool, with an earthen floor and straw-thatched roof. Her entire possessions amounted to her spinning-wheel and other spinning equipment, a few pieces of furniture, an axe and an old spade. Her back garden was planted with potatoes and other vegetables, the front crammed with herbs of every kind. She dispensed herbs to her neighbours, 'with her sage advice and directions as to how they should be used'. Around this little front garden was a hedge of appleringy, lemon balm, sweetbrier, red and white roses (wild and cultivated) and bluebells.[120]

Towards the end of his life the Aberdeenshire weaver, herbalist and self-taught botanist John Duncan (1794–1881) rented a room in a croft house where he was allowed a space for a garden to house his collection of herbs, mostly taken from the wild. His biographer William Jolly described this garden as he saw it not long before Duncan's death:

[119] Bethune, *Poems by the late John Bethune*, 73.

[120] Adam Philip, *The Parish of Longforgan: a sketch of its church and people*, Edinburgh 1895, 274.

> in front of the cottage ... on the sloping bank between it and the burn ... was the large and fertile garden ... . It faced south and was enclosed by a stone wall, having a row of four rowan trees before the door, with bushes along the fence on both sides of it, and willows and geans [wild cherries] at the foot next the burn, planted a little before John came. That part of the garden immediately in front of the [weaver's] workshop was railed off for John's special use, and there he cultivated a large number of plants, chiefly wild, as well as those required for his drugs ... . For years it was kept with the greatest neatness, and without a weed. He spent a good deal of time there, getting up a three o'clock on summer mornings to tend it. Nothing could exceed his delight in working amongst his flowers, especially the wild importations from nature. The natural result of such attention was that his little plot was a treat to be seen, and gained the admiration of all that visited the place.[121]

Ghostly remnants of cottage gardens survive in a corner of Angus. On the site of the vanished settlement of Burnside of Dun near Montrose, once occupied by weavers and labourers, there are patches of ground which are recognisably former gardens. The hamlet was abandoned in the early twentieth century and most of the site has never been cultivated. Today, overgrown beech hedges and gooseberry bushes are still to be seen behind the half-buried outlines of the houses. Jane Anne Duthie knew the hamlet in the 1890s when most of the inhabitants had left, leaving only the very old and infirm. The clay-and-bool houses were falling into ruin, but there was 'a stragglin' rose bush tryin' to force its way through the rubbish; some black currant bushes and honeysuckle and here and there plants that had wont to be used for healin' purposes – "healin' blades" they used to be ca'd'.[122] Those who played amongst the ruins as children later recalled fruit trees and bushes.[123]

Gooseberry and currant bushes, an apple and a pear, culinary and medicinal herbs, climbers and a few flowers that took up very little space: these, together with the essential potherbs and root vegetables, were the basic elements of a cottage garden. But some cottage gardeners, even amongst the poorest, were growing a wider selection of flowers, fruit and vegetables.

At the end of March 1840 in Alexander Bethune's garden the crocuses were going over, auricula and narcissus were opening and lilies pushing up through the earth. All these had been planted the year before by the brothers in their new garden, but John had not lived to see them.[124] In the summer, among the most popular flowers in cottage gardens, and never out of fashion, were creepers and climbers, 'vines', covering the walls and roof. Ivy, honeysuckle, sweetbrier

---

121 William Jolly, *The Life of John Duncan: Scotch weaver and botanist*, London 1883, 321.

122 Jane Anne Duthie, *Rhymes and Reminiscences*, Brechin 1912, 122.

123 Rice, *'All Their Good Friends'*, 90.

124 Bethune, *Memoirs of Alexander Bethune, embracing selections from his correspondence and literary remains*, ed. W. McCombie, Aberdeen 1845, 189.

and other climbing roses – and scarlet runners too – were ideal for gardens where vegetable and fruit cultivation came first, as they took up almost no space and needed little attention once established. They were also the essence of the country cottage in the Romantic style. The home of the weaver poet James Thomson of Kenleith, near Edinburgh, was described by a visitor in 1813: 'The small Gothic windows embowered with honeysuckle and ivy; parterres of roses, and other sweet smelling flowers, encircle its walls.'[125]

Not all the roses were climbers. The horticulturist Frances Jane Hope wrote in the 1870s that among the plants she grew in her flower border in Edinburgh she had the 'old-fashioned and despised Roses [Scots or Burnet Rose] that bear hips, the single white, only to be found in cottage kale-yards (the double variety of which is considered the white rose of Jacobite days)'.[126] However, the hardiest and longest flowering, the China rose, introduced only in the 1790s, was also often seen in these gardens. The burnet rose apart, there is little evidence of Edward Hyam's 'museums of plants' – old varieties 'unwittingly' preserved in cottage gardens in England because of the gardeners' lack of means.[127] If cottage gardens were a relatively recent phenomenon in Scotland, it is perhaps not surprising that very old varieties were missing. A single, tantalising, exception is the strain of matthiola or intermediate stock discovered in a cottage garden in East Lothian, according to Roy Genders.[128] Hyams adds that these flowers had been preserved 'since Tudor [sic] times'[129] but it has proved impossible to find out more about this discovery. 'East Lothian stocks' were available from seed merchants in the 1850s, and are still a popular seed catalogue item.

The 'chosen flowers of the Scottish cottager', said Patrick Neill, were, unsurprisingly, traditional: marigolds, often self-sown, white narcissus, wallflower, London pride and polyanthus ('spinks' in Scotland).[130] In literary evocations of cottage gardens the flowers are securely rooted in the past. Tulips, jasmine, violets, roses, anemones, daisy, primrose and lily appear in the peasant's garden in John Nevay's poem of 1836.[131] An 'old-fashioned garden' in a short story of the kailyard school, contained pinks, daisies, forget-me-not, wallflowers, thyme and moss roses.[132] In real-life gardens, too, it seems that reliable, hardy and easy-to-grow traditional flowers were preferred. In Bervie in Kincardineshire in the 1830s primroses, polyanthus, pinks, rocket, Sweet William, violets, wallflowers

---

125 W. J. I., 'Biographical sketch of Thomson the poet of Kenleith with a desultory description of his Parnassus', *Scots Magazine*, vol. 75, 1813, 645–8, 647.

126 Hope, *Notes and Thoughts*, 99.

127 Hyams, *English Cottage Gardens*, 65.

128 Roy Genders, *The Cottage Garden*, London 1969, 236.

129 Hyams, *English Cottage Gardens*, 65.

130 Neill, *Gardens and Orchards*, 25.

131 John Nevay, *The Peasant: a poem in nine cantos: and other poems*, Edinburgh 1834.

132 Ian Maclaren, 'A wise woman', in *Beside the Bonnie Brier Bush*, Edinburgh 1895.

and tulips grew in the 'little gardens'.[133] On the other hand, competitors even in the early cottage garden flower shows were exhibiting a mixture of traditional flowers such as pinks, hollyhocks and heartsease and quite recent introductions – the brightly coloured bedding plants such as calceolarias, dahlias, French and African marigolds and coleus (see Chapter 7).

Flowers grown in pots on window sills became popular in the last quarter of the nineteenth century and from the flower show reports we know that the pot plants, grown by men and women, included all kinds of pelargoniums (not distinguished then from geraniums), ferns, petunias, begonias and many more. The notorious Victorian partiality for flowerless pot plants like the aspidistra notwithstanding, it seemed brightly coloured flowering plants were preferred by the cottagers.

House walls might have to bear fruit trees as well as ornamental climbers. Feuars' gardens, according to Neill, had apples, pears and cherries sometimes trained against the house wall, along with gooseberry and white and red currants. Neill's recommendation to landowners to plant fruit trees probably indicates their absence from most cottage gardens of the poor. Where they were grown one might expect older varieties to be found, or recommended for them, especially as there is a very old fruit-growing tradition in parts of Scotland. But in fact Patrick Neill's apple recommendations were relatively recent introductions: White Codling, White Hawthornden and Nonsuch. In the Bervie gardens they were also growing the newer apple varieties: the Keswick codling, a recently discovered sweet and juicy cooking apple, and the Ribston pippin, a very popular eighteenth-century variety and probably the forefather of the orange pippin.[134] Crawford, Lemon or Lammas, Green Chisel, Green Yair and Auchan were the varieties of pear Neil proposed.[135] Pears had been grown in Scotland for centuries – they were a mainstay of monastery gardens in the Middle Ages – and some of these varieties were very ancient, if not always good to eat. The moorfowl egg and Jargonelle pear were the Bervie favourites.[136] The Jargonelle was an old established pear dating from the seventeenth century or even earlier, and the moorfowl (Muirfowl) egg was, according to a comment from the early 1800s, 'the best pear of any kind' – hardy, refreshing and juicy. Crawford pears, hardy and ripe in August, were popular, along with Green Yair, well into the twentieth century.[137] The Bervie cherry trees were May Duke and blackheart, both old varieties.

133 *NSA* Bervie, Kincardineshire, vol. XI, 4.

134 Ribston pippin apple, Orange Pippin website, available at https://www.orangepippin.com/varieties/apples/ribston-pippin.

135 Neil, *Gardens and Orchards*, 27.

136 *NSA* Bervie, Kincardineshire, vol. XI, 4.

137 Forbes W. Robertson, 'Pears in Old Scottish Orchards', *Review of Scottish Culture*, vol. 22, 2010, 1–17, 4, 8–9, 12.

Soft fruit was more popular than tree fruit – cheaper to buy, or free if grown from cuttings, easier to grow and prune and quicker to fruit. Gooseberries were the archetypal working-class garden fruit, grown throughout Britain. Prolific, tolerant of poor soil and unfavourable climates, they were also – surprisingly to modern tastes – regarded as sweet treats and delicious when eaten raw. Edward Hyams considered that this fruit was probably indigenous to Britain and domesticated by cottagers.[138] The first English record dates from 1275.[139] When they first appeared in cottage gardens in Scotland is unknown, possibly not before the end of the eighteenth century, as there are suggestions in both the *OSA* and the *NSA* that they were recent introductions. Gooseberry shows, so widespread in the north of England, do not seem to have caught on here and, as far as is known, the fruit was not entered in cottage garden flower shows before the 1850s. They were nevertheless strongly recommended by gardening experts. 'The poor man's friend, and so acceptable to the rich that none are willing to dispense with it. The gooseberry tree may be called the vine of the north', wrote the gardening minister Nathaniel Paterson.[140] Paterson also noted that there were 'almost endless varieties', all of them good. Archibald Gorrie advised ten gooseberry bushes for a labourer's family of six and the Fife gardener Robert Adamson recommended eighteen bushes for a cottage garden.[141]

The story of the gooseberry in Scottish cottage gardens rather neatly symbolises the story of those gardens themselves: late introduction, adapted to difficult growing conditions, a cheap, prolific and hardy bush whose fruit was treasured by gardeners and small boys alike, but not ornamental – and held in little regard today.

Red, black and white currants were also popular bush fruit. Bervie in the 1830s had raspberries and strawberries too,[142] but, from flower show reports at least, these seem not to have become common until mid-century. The removal of the tax on sugar in 1875 meant cheap sugar and a boom in home-made jam. Both garden-grown and wild berries were made into jellies and jams, thus extending the fruit season and encouraging more cultivation of soft fruit, both for sale[143] and in gardens.

Culinary and medicinal herbs were raised in most cottage gardens, as well as being gathered from the wild. Herbal cures survived amongst the rural poor well into the twentieth century, but the range of plants used and ailments treated shrank steadily, as scientific medicine and commercial remedies gained

[138] Hyams, *English Cottage Gardens*, 206–7.

[139] Christopher Stocks, *Forgotten Fruits. A guide to Britain's traditional fruit and vegetables*, London 2008, 98.

[140] Nathaniel Paterson, *The Manse Garden*, Glasgow 1836, 116.

[141] Robert Adamson, *The Cottage Garden*, 2nd edn, Edinburgh 1856, 15.

[142] *NSA* Bervie, Kincardineshire, vol. XI, 4.

[143] Blairgowrie in Perthshire became a successful centre of soft-fruit growing by smallholders in the 1890s, often as a sideline, supplying the Dundee jam factories.

ground.[144] Thyme, spearmint, southernwood, elecampane ('esteemed a sort of panacea'), rue (for ulcers), tansy, horehound (grown in the north for chest infections and later for flavouring sweets) and caraway ('carvey' to add to cheese, shortbread and sweets) could all be found in cottage gardens at the beginning of the nineteenth century.[145] Cottagers often planted chamomile on a little mound, to make a summer seat, said Neill. In the little gardens of Bervie grew 'some of the sweet [culinary] and physical herb seeds, such as marjoram, marigold, rosemary, thyme, etc'.[146]

Nelly Johnstone had marsh mallow, camomile, peppermint and groundsel in her front garden[147] and from other accounts we learn that lavender, tansy, rhubarb (root), hyssop and lemon balm were also grown – and probably many more. These cultivated herbs treated a wide variety of comment ailments, from inflammation and measles (marigold) to minor complaints such as stomach pain (camomile, groundsel), headache (lavender), insect stings (lemon balm), indigestion (peppermint), rheumatic pain (rosemary), coughs (thyme), colds and sore throat (hyssop) and intestinal worms (tansy, groundsel and southernwood).[148] Fife miners were still taking 'vegetable cures' on the eve of the First World War. These seem to have been mainly for external use and did in fact include vegetables: poultices of leek, chewed tobacco leaf, soap and sugar for whitlow; infusions of ivy and kail for eczema and of bramble leaves for diarrhoea – and a potato in the pocket to ward off rheumatism.[149]

The vegetables most commonly found in cottage gardens in the late eighteenth century were kale for the broth pot and potatoes. Kale, the national vegetable, has been grown in Scotland since at least the Middle Ages, and may have been taken there by the Vikings. It is easy to grow and high-yielding, extremely hardy – it stands throughout the winter – and nutritious. George Robertson maintained that it would thrive on the same spot for years.[150] It came in many varieties: Robert McNab in *The North British Cultivator* (1842) wrote about a cottage garden with twenty or thirty varieties of kale of different colours and leaf shapes, including the most popular: Scotch, or Siberian, borecole and German greens, or 'curlies'.[151]

---

[144] David E. Allen and Gabrielle Hatfield, *Medicinal Plants in Folk Tradition: an ethnobotany of Britain & Ireland*, Cambridge 2004, 25.

[145] Neill, *Gardens and Orchards*, 25.

[146] *NSA* Bervie, Kincardineshire, vol. XI, 4.

[147] Philip, *Parish of Longforgan*, 275.

[148] Thomas Bartram, *Bartram's Encyclopaedia of Herbal Medicine*, London 1998, 214, passim.

[149] David Buchan (ed.), *Folk Tradition and Folk Medicine in Scotland, the writings of David Rorie*, Edinburgh 1994.

[150] Robertson, *GV Midlothian*, 116.

[151] Robert McNab, *The North British Cultivator: a treatise on gardening, agriculture and botany*, Perth 1842, 92.

In Forgandenny, Perthshire, in the 1790s cottagers were growing turnips, cabbage, carrots, leeks, onions and peas as well and we have Lord Kames' testimony that some of these were being grown in cottage gardens in the west of this county twenty years earlier. Forgandenny was sheltered, blessed with fertile ('carse') soil and mostly enclosed, indicating improved farming methods were already well established.[152] According to Patrick Neill feuars' gardens grew the same vegetables as the Forgandenny villagers. In village gardens some of the varieties cultivated by weavers and tradesmen were large early, early York and Scotch cabbage (a winter crop), savoys, German greens, red and green curly kale, orange carrots, Strasburg and Deptford onions, yellow and white Dutch turnips, long pod, Lisbon, broad Windsor or Turkey beans, marrowfat peas and Charlton and blue Prussian peas, leeks (Scotch and flag) and chives; occasionally parsley, cress and lettuce, though these were 'not much attended to'.[153] Parsnip,[154] beetroot and cauliflower could also be found but less commonly. Five of these named varieties appear in Philip Miller's *Dictionary of Gardening*, published in 1731. The *Dictionary* tells us that Strasburg onions (now lost) were excellent keepers; the Scotch cabbage was never affected by frost; Dutch turnips and Charlton peas were both early varieties. Windsor beans were grown to be eaten fresh, not dried, unlike most beans and peas, such as the blue Prussian pea.[155] These were also the crops recommended for cottage gardens in Sir John Sinclair's *General Report* (1814) by a Mr Nicol. The vast quantities of seed and plants Nicol lists indicate just how much cottagers were expected to rely on their gardens: 300 to 400 kail plants, 200 cabbages, 1,000 to 2,000 leeks, two to three pounds of bean seeds.[156]

Apart from the usual roots and greens, the vegetables grown in Bervie included salad, lettuce, mustard, cress and radishes as well as other crops not obviously destined for broth such as broccoli, cauliflower and spinach. There are no other accounts of such a wide and daring range so early. Bervie was a prosperous, much improved coastal parish with a mix of farming, fishing and linen-weaving, but whether the variety of crops was unique to the village or whether the author, the Rev. John Glegg, was simply more interested in the contents of cottagers' plots than other *NSA* contributors is impossible to say.

Records of cottage garden shows later in the century reveal that vegetables grown by cottagers for show at least were still almost the same as a century before.

152 *OSA* Forgandenny, Perthshire, vol. III, 298–310.

153 Neill, *Gardens and Orchards*, 21, 24.

154 Parsnips were almost unknown in Scottish greengrocers and supermarkets before the late twentieth century.

155 Philip Miller, *The Gardener's Dictionary*, 9th edn, London 1735, vol. 1.

156 Mr Nicol, 'Directions for the culture of cottage-gardens by the late Mr Nicol', in Sinclair, *General Report of the Agricultural State and Political Circumstances of Scotland*, Edinburgh 1814, vol. II, chapter IX, appendix, article C, 426–7.

Still no garlic, squashes – apart from marrow – or different varieties of peas and beans. Only celery, beetroot and shallots were new.

Not a few cottage gardeners would have shared the conservative attitude and lack of skill and equipment described by John Duncan in his anecdote of his Aberdeenshire neighbours in the 1850s. You can almost hear the tutting. He had 'looked in over yards in country places' as he walked by, 'and seen nothing but a coquiny of green kale and berry bushes, growing like rasp bushes close with weeds'.

> I asked ... some country people what they had growing in their gardens. They said, 'Oh, we have nothing but green kale, but you may come and see our yard. We were thinking to sow a pickle [few] onion ('oin') seed; where would we sow them?' Weel, I looked around me in the yard, and all was close with weeds. They said, 'Oh you will delve a bit and sow the onion seed.' I said to them, 'I have not a spade.' They said, 'We have a spade.' They brought an old spade not above six inches long to me! ... Now they could have a good many useful plants instead of all these weeds: such as horehound and hyssop and sage and caraway and rhubarb and scurvy grass and rue and sweet marjoram and thyme and parsley and parsnips and plenty of green kale, a great many flowers of hardy annuals and biennials and perennials, of various sorts, both for use and beauty.[157]

A bird's-eye view of cottage gardens across the Eastern Lowlands would have shown a patchwork of poorly cultivated patches of kale and potatoes and neatly tended plots bright with flowers and a good variety of vegetables. Which kind of garden prevailed? Conditions in Scotland were harsher than in England: Scottish rural workers were generally poorer than in England and lived in worse housing; the climate was more difficult and soil poorer, outwith some favoured districts. Those who took the trouble to go and look at the gardens, however, were often impressed. Success was very often attributed to the influence of benign landowners prepared to let head gardeners give away seeds and cuttings, run best-kept garden competitions and encourage cottage garden societies. Strict coal owners prepared to fine delinquent miners for neglecting their plots were also lauded. But the efforts of the gardeners were often praised too and it was recognised at the time that gardens could be more productive than fields because the land was worked with the spade rather than the plough, and because more manure was applied to it.[158]

As with the farm servants' kailyards, East Lothian, Fife and Perthshire seem to have had the best cottage gardens throughout the nineteenth century and the reasons are obvious. East Lothian was the most prosperous agricultural

157 Jolly, *Life of John Duncan*, 267.

158 Allotment crop yields in England could be nearly double those from farms (Burchardt, *Allotment Movement*, 156). HASS promoted spade work for cottagers with premiums, offered from 1841.

Plate 18 Mr Melrose in the Station Garden, Currie, Midlothian, no date.

county in Scotland, with the best soil and climate, and rural workers were the best paid. The exceptional number of cottage garden societies in Fife, the high rate of resident landowners and the climate all played their part in the county's enthusiasm for cottage gardening. The climate was judged 'extremely favourable for the growth of fruit, and the crop of fruit of all descriptions which some of [the] neat little gardens produce in a year is perfectly marvellous'.[159] In Perthshire, too, all these factors – good soil, well-built cottages and a strong horticultural tradition – came into play.

Rising standards of living, greater security of tenure and a sense of independence were at least as responsible for better gardens as the benevolent or otherwise authority of the landed proprietors and employers. In the Fife mining village of Kelty, for example, the houses the miners owned contrasted strongly with those the company owned: 'They take pride in their home, and tending the garden becomes a supreme pleasure. The miner's wife seizes the opportunity to give rein to her womanly instincts which appear when she is left free to make the home in her way. In short, it raises the whole standard of life'.[160]

The story of the cottage gardens worked by a wide spectrum of the rural working class, both work and home balanced on the whim of employers and

[159] Macdonald, 'On the agriculture of the county of Fife', 34.

[160] Kellogg Durland, *Among the Fife Miners*, 1904, available at http://www.scottishmining.co.uk/461.html.

landowners, to the beginnings of rising comfort and security is also the story of a slowly emerging gardening culture. Those same employers and landowners clearly understood the welfare and moral significance of the gardens, and those not driven, like the coal owners and the NLB, to bully their workers into gardening made efforts at encouragement. But it was the enthusiasm and commitment of the gardeners themselves that effected the transformation of so many villages and cottages, albeit slowly and haltingly (Plate 18).

CHAPTER 4

POTATO GROUNDS

Rejoice, ye bodies, ane and a'
Lang hae your backs been at the wa';
But ye may clap your wings and craw,
Nae mair to grieve,
There's taties now at ilka shaw
As big's your nieve. [fist]
...
For taties mak a gusty meal
Whan they are chappit up wi' kail
An half a pound o' butter hale -
Flung in the pat;
A hungry chield wi' that I'll bail,
Will no' find fau't.
Be glad, ye poor throughout the land,
That siccan plenty's come to hand,
The markets high will get a stand,
They'll all be down,
And disappoint the wicked band
O' half a crown.[1]

James Thomson, the weaver poet of Kenleith in Midlothian, wrote this poem in 1796 in expectation of a good potato crop. The following year the minister of lowland Bendochy in Perthshire wrote, 'The potatoe is the true root of Scarcity, which promises to set Famine at defiance.'[2] The failure of the 1794 harvest, grain hoarding by speculators, the activities of English agents and fear of bringing produce to markets on the part of farmers had all led to rocketing prices and

[1] James Thomson, weaver in Kenleith, 'Potatoes', *Poems in the Scottish Dialect*, Edinburgh 1801.
[2] *OSA* Bendochy, Perthshire, vol. XIX, 351.

outbreaks of violent protest. There were over thirty food riots in towns and villages from Angus to the Borders in the late 1790s.[3]

The minister of Bendochy was writing between the two grain crises that led to these riots – 1794–6 and 1799–1801 – and it was the first of these that prompted the newly constituted Board of Agriculture to promote potato cultivation to feed the poor. Sir John Sinclair, its first president, took cautious credit for the Board having – probably – been 'the happy instrument of preventing the risk of scarcity and famine in the ensuing season'.[4]

Potatoes were first grown in the gardens of landowners in Scotland in the late 1600s and soon after in those of larger tenant farmers. There are at least three stories about their subsequent adoption as a field crop. According to one account, it was the factor of an estate in Kilsyth who experimented successfully in 1739, and then demonstrated the method to agriculturists in Dundee, Perth and Edinburgh. Another version has Thomas Prentice, a labourer in the same parish, planting potatoes on his plot in 1728 and being copied by his neighbours.[5] Prentice made a fortune from his potato crop, it was said.[6] George Robertson has a third tale of potatoes being first introduced in the village of Marykirk, Kincardineshire, by an old soldier. But as he lived there for only a year and the villagers never learned to grow them potatoes were not cultivated in that county until the 1760s.[7] After 1750 potatoes slowly became a field crop throughout the Lowlands, but cultivation did not really take off before the 1780s, probably stimulated by the failure of the grain harvest in 1782.[8]

According to the great historian of the potato, R. C. Salaman, potato growing was taken up faster by the poor than by the farmers.[9] No other foodstuff, not even oatmeal, was so vital to the daily life of the poor. And it was essentially *their* food. Salaman, points out that, unlike the Highlands and Islands, where lairds introduced it, in the Lowlands it 'established itself partly because it was admirably suited to a peasantry, rapidly passing from the stage of small communal holders into landless labourers' as well as suiting the needs of the new rotations of improved agriculture.[10]

In 1795 the Board of Agriculture's committee appointed to investigate solutions to the food crisis reported that 'among the various ideas which were with the most patriotic views expressed on that occasion, none seemed to merit so immediate an

3 Catherine M. Burns, 'Industrial Labour and Radical Movements in Scotland in the 1790s', unpublished MSc thesis, University of Strathclyde 1971, 98–100.

4 'Substance of Sir John Sinclair's Addrese to the Board of Agriculture', *Scots Magazine*, 1st August 1795, 22–5, 24.

5 Ibid., 390.

6 *Chambers Edinburgh Journal*, 1835, vol. III, 119.

7 Robertson, *GV Kincardineshire*, 222.

8 Flinn, *Scottish Population History*, 423.

9 R. N. Salaman, *History and Social Influence of the Potato*, rev. edn, Cambridge 1985, 408.

10 Ibid., 392.

attention, as to encourage, by such means as were in the power of the Board, a more extensive Cultivation of Potatoes'.[11] Drawing on the county agriculture reports (the *GVs*) published the year before, the report, amongst detailed though frustratingly unscientific information on varieties, manure and methods of cultivation, made one clear and wholehearted recommendation – that every gentleman of landed property should provide their cottagers with attached land to grow potatoes. On average, 128 perches (approximately three quarters of an English acre) would produce 240 bushels of potatoes (nearly six tons), which would feed a family of five for a year and to spare – monotony was not a consideration. Half that amount of land would yield enough to supplement their diet and be 'highly advantageous'. But they cautioned that the family would need extra land to keep a cow and a pig to produce the manure the crop would demand.[12]

Potato grounds were not invented by the Board of Agriculture. They had probably been first introduced in Leicestershire, where farmers gave their workers permission to plant potatoes in the corners of fields – 'they plant and clean in mornings and evenings at their own time.'[13] The beauty of this, the committee noted, was that not only did it benefit the poor but they were 'thus indulged [sic] without any expense to the farmer, who grants so commendable a favour'. The report was not able to identify a benefit to the farmer in terms of improved rotations, however, and there was a (unfounded) suspicion that potato crops exhausted the soil. On the other hand, having workers bring wasteland or moors or mountainsides into cultivation by raising potatoes in their own time could only be advantageous to the farmer.

The practice of granting potato grounds was quickly taken up in the Highlands but established more gradually in the Eastern Lowlands – initially in East Lothian and spreading around the south-east.[14] Grounds were strips of land – rigs or drills – provided by the farmer either on uncultivated land or on his fields both for his own servants, who had them as part of their wage, and for labourers and artisans, for rent. Workers held them for the duration of the potato-growing season, around six months, after which the land was returned to the farmer. Each year, rigs on different fields would be offered. The usual terms were for the farmer to do all the work that required horses – ploughing and carting – and for the workers to provide the seed, tend the crop and lift it. The farmers did not always get the best of the bargain, though, sometimes ending up with a patchwork of well and badly cultivated rigs.[15] In addition, on

11 Board of Agriculture, *Report of the Committee of the Board of Agriculture appointed to extract information from the County Reports and other Authorities concerning the Culture and Use of Potatoes*, 1795, v, available at https://play.google.com/store/books.

12 Ibid., 72.

13 Ibid., 11.

14 George Buchan-Hepburn, *GV East Lothian*, Edinburgh 1794, 92; Douglas, *GV Roxburgh*, 133.

15 Douglas, *GV Roxburgh*, 134.

Plate 19 Potato grounds at Pennan, Banffshire, 2005. Traces of the grounds can be seen to the right of the harbour.

farms near towns and villages farmers could let small plots for potatoes (from one eighth to one sixteenth of an acre) to cottagers and – very often – to fisher families. The fisherfolk of St Monans, Fife, paid 2s. per 100 yards in the 1830s for their potato grounds – nearly double the average rent in the parish. If labour and manure was provided by the farmer, the rate went up to 4s. In East Lothian rents were as much as five guineas per acre.[16] In Berwickshire potato grounds were most often to be found near towns and larger villages, where, according to George Robertson, 'the mechanics themselves are the chief cultivators'. They rented between an eighth and a tenth of an acre and raised between sixteen hundredweight and one ton of potatoes.[17] At the fishing village of Pennan on the Moray Coast the outlines of a considerable number of potato grounds can still be seen on rising ground to the east of the harbour (Plate 19).[18]

Jeremy Burchardt argues that in England it was labour shortages during wartime harvests that prompted the offering of potato grounds.[19] Labour shortages, indicated by rising wages, were seen in Scotland too, as alternative employment in towns drew young men off the land. However, the granting of grounds as part of the long hire wage suggests that they played a different role in Scotland: only in Roxburghshire is there report of grounds being offered in exchange for seasonal work,[20] while – unlike in England – they seem to have become more valued by farm servants over time. Scotland's climate was ideal for potato cultivation.

16 Buchan-Hepburn, *GV East Lothian*, 105.

17 Robertson, *Rural Recollections*, 319.

18 Personal communication from D. G. Lockhart, 20th September 2018.

19 Jeremy Burchardt, 'Land and the laborer: potato grounds and allotments in nineteenth-century southern England', *Agricultural History*, vol. 74, no. 3, 2000, 667–84, 671.

20 Douglas, *GV Roxburgh*, 133.

While English potato grounds were found mainly in the West of England, where the mild, wet climate and peaty soils offered ideal growing conditions, the whole of Scotland (like Ireland) offered such conditions.

The Board of Agriculture's recommendation, which they considered eminently practicable, of three quarters of an imperial acre per family[21] was never adopted, at least not in Scotland. Even half that quantity, which the Board accepted as 'highly advantageous', was unusual.[22] In 1814 East Lothian farm servants were allowed 100 yards of a rig nineteen and a half feet wide, or 900 yards of a drill. The farmer dunged it and his horses ploughed it, after which the wife and children weeded. 'The importance of this privilege is too obvious to be insisted upon.'[23]

The potato ground habit spread slowly up the east coast and by the 1830s was becoming general, although in the *NSA* references to the grounds are found more often in south-eastern and Fife parish accounts than elsewhere. Potato grounds increasingly made up part of the payment in kind for farm servants. The amount of land allocated to hinds varied between counties and between farms – between 1,000 and 1,500 linear yards in the south-east at this time, less in Fife. Usually the grounds took the form of a rig in a field, but sometimes they were patches of marginal or waste ground. Potatoes could be, and were, grown everywhere, however – in corners and odd little patches – which was another advantage over cereals such as oats. Alexander Somerville recalled his father working up to four hours a night by the light of the moon raising potatoes in tiny strips, no more than eighteen inches wide, alongside the farm dykes.[24] His father later convinced the farmer that the thorn hedges would benefit from having potatoes grown alongside them, and the practice spread to other farms.

The custom of horse work by the farmer and hoeing and earthing up and harvesting by the worker and his family continued. Sometimes the farmer also manured the rigs and supplied the seed potatoes, but it was always the servant and his family who lifted the crop, which was then carted home by the farmer. In East Lothian the 'cottar dung' system prevailed: in addition to supplying the seed potatoes and lifting the crop the cottar (hind) collected dung for a year and his master carted it to a piece of ground – 'probably not the richest in the farm', noted the farmer George Buchan-Hepburn – and spread it at a rate of fifty cartloads per acre. The farmer 'led home' the crop to the cottage.[25]

Potato grounds served both farmers and their servants well. One and a quarter tons or thereabouts was the usual yield in Berwickshire in the 1820s, enough to feed not only the family through the winter but also the resident pig.[26]Potatoes

---

21 Board of Agriculture, *Report on Culture and Use of Potatoes*, 72.

22 The manure needed for such a large ground might have made this unfeasible.

23 Sinclair, *Account of the system of husbandry*, 132.

24 A. Somerville, *The Whistler and the Plough*, 214.

25 Buchan-Hepburn, *GV East Lothian*, 92.

26 'Verus', 'On field gardening', 260.

were a cheap, sustaining and storable food for the poor but, just as importantly, a potato crop was soon found to prepare the ground for a cereal ('white') crop the following year. The cottar dung system made for heavier manuring than the farmer had time or dung for, while potatoes took different nutrients from the soil than cereals. The potato crop was weeded more easily and more thoroughly than other crops – potatoes thus 'cleaned' the soil.[27] So potato grounds supported an effective crop rotation with minimal effort on the part of the farmer, as well as bringing waste ground into cultivation.

Potato grounds, mainly as part of the farm servants' payment in kind, kept their place across most of the Eastern Lowlands at least until the 1880s. In England, meanwhile, the popularity of potato grounds peaked in the 1830s and then gave way to allotments, which could be rented long-term and where the labourer's efforts benefited only himself and his family.[28] By the 1860s they were common in only a handful of counties in the south and east of England, and had disappeared in most of the south. It was mainly stockmen and dairymen who took them: hired annually, working long hours and often living on the farm, just like the ploughmen of Scotland.[29]

Table 1 Percentages of farm servants with potato grounds, by county.[30]

North-east	Percentage of farm workers holding potato grounds	Central	Percentage of farm workers holding potato grounds	South-east and Borders	Percentage of farm workers holding potato grounds
Nairn	52	Forfar	9.7	Haddington	39
Moray	23	Perth	9	Peeblesshire	?
Banff	9.9	Kinross	0	Selkirkshire	49
Aberdeenshire	9.9	Clackmannan	0	Roxburgh	74
Kincardine	9	Fife	26	Berwick	62
		Edinburgh	3		
		Linlithgow	2		

As Table 1 shows, by the 1880s it was in the most rural counties, in the Borders and beside the Moray Firth, that potato grounds were most common. In the Central Belt, apart from Fife (always an outlier), few farm workers bothered

27 Burchardt, 'Land and the Laborer', 672.

28 Ibid., 669–70.

29 Ibid., 674.

30 *Return of Allotments*, Table II, 9.

with potato grounds, the men preferring to take potatoes 'from the barn floor' or the pit.[31] Evidence from the Royal Commission on Labour shows that potato allowances for farm servants were typically twelve or sixteen hundredweight and were sometimes unlimited. Where potato grounds survived, the servants often sold the bulk of their crop. Many farmers were happy to give up the grounds, but the Assistant Commissioner for the 1893 Royal Commission on Labour, R. Hunter Pringle, who strongly favoured the retention of payment in kind, argued that the yield of the grounds was worth to the men three times the cost to the farmer.[32] Pringle also investigated the disparity in drill (dreel) length between two neighbouring farms in East Lothian: they could vary in length from 500 to 1,000 or even 1,600 yards. It turned out that the farm where 1,600 yards was allocated was on soil less suitable for potatoes, which was dressed with less manure, and the drills were narrower. Potatoes from this farm sold for 50s. per ton, compared to £4 from the other farm, with shorter drills and higher yields. He also discovered that servants around Dunbar – the most fertile part of East Lothian – were in the habit of selling half their crop of prize 'Dunbar reds' to the masters and buying cheaper ware for themselves.[33]

In the pastoral districts of the Borders potato grounds were still part of the wage paid to shepherds in the 1890s.[34] In the memoirs of the Borders sheep farmer Robert Shirra Gibb they are identified as the only part of the 'gains' system that had not died out by the 1920s. According to his account, the survival was entirely due to the generosity of the farmer. Following the introduction of paid holidays, the worker's responsibility was reduced to providing the seed. The servant then took three days' holiday: one for planting, one for hoeing and one at 'howking' time. 'So in the end, the farmer did all the work.'[35] The servant had their masters over a barrel: the 'fallow brake'[36] allotted to the servant was 'part of the contract', as one northern farmer put it: 'I will not get a labourer unless I give the potato ground', he declared, even though he lost money by it; and when the ground was returned it was covered with weeds from the poor-quality manure.[37]

Although payment in kind was breaking down everywhere, several Aberdeenshire farmers giving evidence to the Royal Commission on Labour in 1893 said they had tried to pay 'all money' wages but without success: servants still wanted their fuel, their potatoes and their oatmeal. These gains may have

31 *RC Labour*, vol. III, pt II, Fife, Kinross and Clackmannan, appendix D, 85.

32 Ibid., Edinburgh and Haddington, appendix D, 66.

33 *RC Labour*, vol. III, pt II, Edinburgh and Haddington, 107.

34 *RC Labour*, vol. III, pt II, Selkirk and Peebles, 190.

35 Shirra Gibb, *A Farmer's 50 Years*, 22–3, 34.

36 The farmer's fields lay fallow one year out of five and when fallow would be given to the servants as potato ground.

37 *Report SC Smallholdings*, 424.

given them a feeling of greater security: they could either consume the potatoes and oatmeal at home or sell them on for cash.[38]

'I have heard it stated that in many cases', wrote Pringle, 'fair play is not dealt out to the hinds in the matter of the land set apart for their potatoes. They are said to be put into the headlands, or shaded places out of the sun, or upon a spot known to be bad soil.' Pringle allowed there may have been such cases but he knew of farmers who let their servants choose their 1,000 yards in any part of any field: 'I am confident in my own mind that, in this matter, strict justice (in many cases uncalled-for liberality) is displayed by the majority of farmers.'[39] In Fife, he reported, the master hoed the servants' potato drills, to save time, and also sold them new seed potatoes cheaply from his own stock.[40]

As in the garden, and on the farm too, working on the potato grounds was a family affair. Women carried out much of the work. Children too were involved in the raising of potatoes, just as they were in the cottage gardens. It was the children who scraped the dung off the road to spread on the potato grounds, who weeded the dreels and who helped with the howking in the autumn. They also worked for the farmer. Elementary education was made compulsory by the Education (Scotland) Act of 1872, which threatened to upset the potato economy of the east coast. The Scotch Education Department fought hard but mostly in vain against the pernicious use of child labour during the potato harvest in October and November. Schoolmasters coping with the seasonal catastrophes that kept country children off school – scarlet fever, snow, and so on – found this the most disruptive event in school life. Children as young as five left the schoolroom to earn extra money with their mothers, lifting potatoes for weeks at a time.[41] Many were taken out of school for planting in April too. In potato-growing areas – most of the Eastern Lowlands – battles between schoolteachers, inspectors and the Scotch Education Department on one side and children, their parents and farmers on the other ensued. Although each parish School Board could set its own holidays without altering the total number of holidays – six weeks a year – the Department awarded grants on the basis of attendance. At Dun school in Angus, a prime potato-growing county, more than half the children between five and thirteen were absent on 6th November 1875. In 1889 eighty per cent were off in early October. The School Board, made up of landowners and farmers, was complicit, lifting compulsory attendance in the autumn in one year and instituting a three-week holiday for tattie-howking in 1893.[42] Later the summer holiday was shortened in rural areas to allow for a two-week tattie holiday in October. The money earned in long days of backbreaking work paid for

38 *RC Labour*, vol. III, pt I, Aberdeen and Kincardine, 135.

39 *RC Labour*, vol. III, pt II, Edinburgh and Haddington, 108.

40 Ibid., Fife, Kinross and Clackmannan, 62.

41 Rice, *'All Their Good Friends'*, 73.

42 Ibid., 73.

winter clothes. Child labour in the tattie fields came to an end in the 1970s, but to this day a two-week tattie holiday survives in Angus schools.

POTATO CONSUMPTION AND CULTIVATION

In normal years potatoes came to fill the 'hungry gap' between the last of the previous year's grain and the next harvest. But for the poor they became the main food. The *GV Fife*, published in 1800, reckoned that they provided one third of the food of the common people for eight months of the year.[43] Several parish reports in the *OSA* refer to the beneficial effects of potato growing for the poor: in Kilconquhar, Fife, 'potatoes were scarcely known ... 40 years ago ... they now afford the poor half their sustenance';[44] in Bellie, Elgin, 'the potatoe is much cultivated by those [with] ... small spots of ground, and proves very useful';[45] and in Ceres, Fife, 'Potatoes form a considerable part of the food of the people and are also used as food for horses.'[46] The acreage planted with maincrop potatoes increased rapidly from the 1790s as urban demand grew: in 1800 6,000 Scotch acres (approximately 7,500 imperial acres) in Fife fed livestock, rural workers and towns and villages.[47] Potatoes were easy to raise and, as T. M. Devine has pointed out, enabled 'a dramatic increase in food supply without the need for radical change in traditional methods of cultivation, technology or social organisation'.[48] Although peck for peck they are less nutritious than grain,[49] the yields were so much greater – three to five times – per acre that it is not surprising that potatoes were adopted so quickly in Ireland and the poorest parts of the Highlands, and by the Lowland labourers. Potatoes can go straight from the drill to the pot without processing or added expense and can, if necessary, provide a complete diet by themselves.

At the beginning of the period, cottagers, like farmers, were mostly growing maincrop or late potatoes. Earlies were grown, according to a report by Patrick Neill and Dr Duncan to the Highland and Agricultural Society in 1814, only by gentlemen in their gardens, and most cottagers outwith the Lothians were apparently unaware of them at this time. This report was intended to remedy this: having a crop of potatoes in the summer months when oatmeal prices were

43 Cited in Salaman, *History of the Potato*, 394.

44 *OSA* Kilconquhar, Fife, vol. IX, 289.

45 *OSA* Bellie, Elgin, vol. XIV, 266.

46 *OSA* Ceres, Fife, vol. V, 382.

47 *GV Fife* 1800, cited in Salaman, *History of the Potato*, 394.

48 Devine, *The Great Highland Famine*, Edinburgh 2004, 13.

49 In his 1869 report on the diet of labourers in Scotland, Robert Hutchison argued that potatoes were far less nutritious than oatmeal (Robert Hutchison, 'Report on the dietaries of the Scotch agricultural labourer', *THASS*, 1869, 1–29, 5–6). Compared with oatmeal, potatoes are in fact lower in protein, though high in vitamin C and potassium.

at their highest would be a great saving (of 4s. a week for six weeks), especially as earlies can be sown closer together than maincrop, so the yield is almost the same. And, as the crop was lifted, the ground could be planted with greens and leeks – this was already the practice in the Lothians. The two main problems Duncan and Neill foresaw were the cottagers' difficulty in acquiring seed, and the question of how the word could be spread. They had a solution very typical of the time: gentlemen landowners should 'encourage the planting of earlies and ... put it in the power of the neighbouring cottagers to procure seed-stock of some of the best early varieties'.[50] This strategy – or perhaps it was the efforts of the cottagers themselves – seems to have led to widespread cultivation of earlies.

By the time of the *NSA*, in the 1830s, potatoes were popular nationwide and had been adopted as a staple of the diet by all workers, not just the poor. Potato consumption then kept rising. Potatoes were, however, not a perfectly behaved crop. Most varieties could not be stored for more than a few months and so were mainly a winter–spring food. Before the potato blight of the 1840s, varieties of potato tended to be regional and short-lived, losing their vigour after about ten to thirty years and often succumbing to viral disease. An exception was the ashleaf, a good early variety, recommended by authorities as distinguished as Arthur Young and Sir John Sinclair and grown throughout the century.[51] The blight of the 1840s devastated Lowland crops as completely as in the Highlands. However, because of the lesser dependence on potatoes than in the Highlands and the more varied diet of the people, including the poor, in the Lowlands the social impact of the crop failure was far less. Extreme hunger and large-scale emigration did not afflict the Lowlands.[52]

The blight – *Phytophthora infestans* – affected all the existing varieties equally and stimulated a renewed interest in selection and breeding.[53] Nevertheless, it was not until the 1860s that efforts by skilled breeders, often Scots, made an impact – for example with the hugely popular Victoria and Champion varieties from William Paterson of Dundee and John Nicoll of Fife respectively.[54] These were exhibited in many cottage garden shows alongside many others. From the 1880s shows often had separate classes for different potato varieties: ashleaf, kidney and round, both red and white, were popular. The Cornhill, Aberdeenshire show of 1911 had seven potato classes: the Champion, forty-fold (a blue-skinned tuber, possibly related to Kerr's Pink), the Kepplestone kidney (a Scottish variety, also blue and still available today) and the Rosebery, in addition to the ashleaf varieties.[55] According to the Royal Commission on Labour, farm servants

50 Dr Duncan and P. Neill, 'Report to the Highland Society respecting early potatoes', in Sinclair, *General Report*, chapter IX, appendix vol. II (A).

51 Salaman, *History of the Potato*, 163.

52 Flinn, *Scottish Population History*, 433.

53 Ibid., 165.

54 Ibid., 167, 168.

55 *Aberdeen Journal*, 11th August 1911, 4.

in Kinross-shire and Clackmannanshire in the 1890s were growing the early varieties ashleaf kidneys, American early rose, Beauty of Hebron and Flounders.[56]

Scotland today has nearly one quarter of the potato-growing acreage of the UK, largely in the east and above all in Angus. Although there has been a steady decline in the twenty-first century, consumption of potatoes and potato products (mainly chips and crisps) is still high and potatoes are a much-loved comfort food.[57] At around 100 kilos per person per year, however, this is a far cry from the more than 282kg of potatoes per person per year served in Berwickshire farm cottages in the 1820s.[58] At such a rate of consumption (probably declining through the century but still much higher than today on the eve of the First World War), the huge importance of this crop and of the potato grounds becomes evident. Self-sufficiency in this nutritious food undoubtedly helped to maintain the health and well-being of rural workers at a level well above that of their urban counterparts.

56 *RC Labour*, vol. III, pt. II, Fife, Kinross and Clackmannan, 61.

57 428 https://www.potatopro.com/united-kingdom/potato-statistics, accessed 2 June 2021.

58 Counting the pig as an adult.

CHAPTER 5

THE MIDDEN

'The midden is the mither o' the meal-kist'(old Scots proverb)

The midden was essential to every garden. The story of the cottage midden in this period is the story of the gradual separation of household waste from the dwelling and its inhabitants. Everyone had a midden in the eighteenth century (and before), except in the most crowded cities where, in theory, muck was cleared from the streets every night. All refuse – animal, household and human – was piled or poured onto the midden. After Archibald Grant of Monymusk, a celebrated improver, had evicted his cottars, he instructed his factor to dump the houses on farm dung hills.[1] Eventually, it was concern not only for health but also for appearance that drove the clean-up. But the continuing need of both farmers and cottagers for spreadable muck slowed it down. There were middens, though fewer, throughout the countryside on the eve of the First World War, and much of the same kind of household waste was being used as fertiliser on many fields and gardens in the twentieth as in the eighteenth century.

The size of the midden and the amount of manure the cottager could collect also determined how much land it was practicable to cultivate as a garden or potato ground. A garden without a source of manure was worthless and, as we have seen, payment in kind for farm servants always included manure for the garden; many farmers also provided the manure for the potato grounds. Potatoes were an especially demanding crop – up to forty tons of manure per acre was recommended in one *THASS* article.[2] Archibald Gorrie, in his master plan for the cottage gardener, stipulated four barrow-loads of pig manure for two falls (seventy-two square yards) of early ashleaf potatoes. Different crops had different requirements: pig manure mixed with green weeds would produce nice floury maincrop potatoes; onions and leeks did best with a mixture of soot

1 Smout, *History of the Scottish People*, 294–5.

2 James Bell, 'Account of the culture and produce of a field of potatoes in the vicinity of Leith', *THASS*, 1803, 471–8, 471.

and poultry dung; early cabbage and other greens with liquid manure composed of soap-suds, 'chamber lie' and any dish-washings not wanted for the pig.[3]

Rules about manure make it clear how important a sufficient supply was. Newly arrived lighthouse keepers were obliged to compensate their outgoing colleagues for it;[4] in the Carse of Gowrie the practice was for the departing hind to sell his leftover manure to the master and in *The Clodpole* the Carse of Gowrie ploughman complained bitterly that this meant the incoming worker was left with less than half a year's supply before seed-time. This may seem, he admits, 'a mere bagatelle' to his readers, but an extra two cartloads of manure would produce at least four more bolls of potatoes for the family and their pig.[5] In the 1890s a group of farm servants, all working on the same farm in Banffshire, told the Royal Commission on Labour Assistant Commissioner that when they left this hire they were not compensated for manure they had spread on their gardens.[6]

MANURES BEFORE IMPROVEMENT

The importance of manure had been known both by scholars and by farmers for thousands of years. Roman writers on agriculture such as Verro and Columella discussed manure in the context of the ancient concept of the four elements of fire, air, water and earth. Their writings remained influential until the Enlightenment.[7] In the Middle Ages its importance was well understood by landowners and peasants alike: dung spreading was one of the tenant's obligatory duties for the lord. In the late Middle Ages agricultural treatises advised on how to prepare manure, by, for example, spreading straw and bracken in yards and on roads to be trampled by cattle and other livestock to speed up decomposition. They also gave the best times of year for manuring and recommended the ploughing in of stubble.[8] By this time more scientifically minded farmers were separating types of manure into different heaps – dung, vegetable waste, animal bones and ashes. Poorer farmers would only have a single heap, onto which not only these items would be thrown but also broken pottery (allowing archaeologists to date manuring).

But it is the discovery of plaggen soils in sites across north-west Europe that confirms most conclusively the ancient use and understanding of the dung heap.

3 A. Gorrie, 'Essay on cottage gardens', 65–6.

4 NRS, NLC, Northern Lighthouse Board, Instructions to light-keepers, 1867 edn, chapter XVI, 3.

5 *The Clodpole*, 8.

6 *RC Labour*, vol. III, pt 1, Moray, Banff and Nairn, 124.

7 Richard Jones, 'Why manure matters', in Richard Jones (ed.), *Manure Matters – historical, archaeological and ethnographic perspectives*, Farnham 2012, 2, 8.

8 Jones, 'Understanding medieval manure' in Jones (ed.), *Manure Matters*, 147–8.

These sites were first identified in the Netherlands but have since been found in the Northern Isles and in many northern European countries. The layers of black or dark-brown organic material, deposited from at least the Middle Ages up to the twentieth century, were constructed from turves used as animal bedding and building material, with the addition of seaweed, cattle dung and household waste, and laid on the infield.[9] The term 'anthropogenic soils' reflects more recent excavations where a wider variety of ingredients has been found. Depending on where they are, these artificial soils can contain sand, charcoal, seaweed, manure and earthenware fragments. As a result, they differ in colour and vary in depth up to one and a half metres. Ben Pears writes: 'Despite this observable variability each site exhibits a deepening around the settlement centres which illustrates that the central kaleyard or garden areas were the primary foci for the deposition of domestic settlement waste including food, organic detritus, carbonized particles and industrial detritus.'[10] Sand and shells could be added to counteract the acidity of peat, turf and dung. It is, of course, no accident that both the midden and the kailyard, as well as the infield (known in some places as the 'muckit land'), were closest to the heart of the settlement, both being essential for human and animal life and needing constant attention.[11]

Everywhere, the farm midden was cleared away every spring and dumped on the infield land, leaving a stinking puddle behind. In the eighteenth century the midden became vital to the success not only of the harvest but of the improvement project. New rotations, new crops such as potatoes and turnips and waste land as it was brought into cultivation demanded ever more creels or cartloads of manure, conveyed to the fields by women or horses.[12] It was recognised that animal and poultry dung was the most effective but household refuse came next in value. One of the more compelling arguments for feuing new villages was that they could provide a constant and convenient supply of manure for neighbouring farmers, and the design of new villages incorporated this: off the broad ruler-straight streets, at regular intervals, paths ran back at right angles to the ends of the gardens so that refuse could be conveyed to the dung heaps and the heaps then carted out to the street and to the farmyards. More than one of the *General Views* remarked that farms near towns and large villages had access to the greatest supplies of manure and could adapt their

9 Smout, *Nature Contested: environmental history in Scotland and Northern England since 1600*, Edinburgh 2000, 70.

10 Ben Pears, 'The formation of anthropogenic soils across three marginal landscapes on Fair Isle and in the Netherlands and Ireland', in Jones (ed.), *Manure Matters*, Farnham 2012, 113.

11 Ibid., 113–14.

12 John Shaw, 'Manuring and Fertilising the Lowlands 1650–1850', in S. Foster and T. C. Smout (eds), *The History of Soils and Field Systems*, Aberdeen 1994, 113.

rotations accordingly, while those further from such settlements needed to include fallow.[13]

IMPROVING MANURES

The first book on manures in English based on scientific discoveries was not published until 1796, and practical advice based on experience and on experimentation was taking over from classical theory in the agricultural literature. The *General Views* are full of observations and discussion of the relative value of different manures for different crops – lime and marl, town muck, woollen rags, malt dust, peat ash, rabbit dung, salt, soapers' ashes and animal bones.[14] John Thomson, in the *GV Fife*, gives a recipe for compost dung hills:

> This is a good plan for furnishing a supply of manure, when a sufficiency of animal dung cannot be had. They consist of earth, scourings of ditches, road-soil, peat-moss, mossy earth, lime, and dung, and such other kinds of vegetable substances as can be easily got, and otherwise of little use, such as quick roots [couch-grass] gathered from the land, ferns, leaves of trees etc. These are all mixed and turn over repeatedly, and suffered to ly and rot for several months, perhaps a year, before the compost is laid upon the land.[15]

Seaweed, or seaware, utilised since at least the Middle Ages, was still recommended. With the same nitrogen content as farmyard manure and twice the potassium, although only one third the phosphate, it was especially valued for bere crops and was effective for up to two seasons. Little wonder that after a storm country people rushed to the seashore to collect as much as possible – a much easier way to harvest than by cutting from the rocks with a sickle. Ware roads can be found throughout Scotland, like the one in East Lothian, which meanders through the Tyninghame estate to Belhaven Bay.[16]

Untreated or semi-composted human excrement was commended by gardening authorities, agricultural writers and landowners as providing unparalleled fertiliser and in Scotland human waste was much more likely to be applied to fields than in England.[17] The *OSA* reports for parishes around Edinburgh reveal that farmers used night-soil collected from the streets of the capital to such an extent that they were able to raise more wheat crops, with less fallow,

---

13 James Robertson, *GV Perth*, Edinburgh 1813, 85.

14 Donald Woodward, '"Gooding the earth": manuring practices in Britain 1500–1800', in S. Foster and T. C. Smout (eds), *The history of soils and field systems*, 104.

15 Thomson, *GV Fife*, 247.

16 Shaw, 'Manuring and fertilising the Lowlands', 116.

17 D. A. Davidson and T. C. Smout, 'Soil change in Scotland: the legacy of past land improvement processes', in A. G. Taylor, J. E. Gordon and M. B. Usher (eds), *Soils, Sustainability and the Natural Heritage*, Edinburgh 1996, 50.

than elsewhere and the price of town muck rose from 2d to 1s. 6d per cartload.[18] Liquid manure – diluted sewage – was enthusiastically adopted from the 1840s following successful experiments by an Essex farmer who sprayed it on his fields using a steam-powered hose. It was found to contain valuable quantities of nitrogen, potash and phosphates, was cheaper than imported guano and was sold by towns and cities in the enormous quantities needed by increasingly intensive farming methods. Chemists were sceptical as to its value as fertiliser,[19] but Lothian farmers near Edinburgh remained enthusiasts for liquid manure and eventually 300 acres to the east of the city was covered with irrigation channels. Four to five crops per year could be taken, and as a result rents nearly doubled.[20] The inhabitants objected strongly to this practice and Edwin Chadwick admitted it was 'probably' harmful to health. He believed nevertheless that diluting the sewage with the water of the Foul Burn reduced the poisonous gases (miasmas), still held at this time to be the causes of disease. By the time of the 1865 Sanitary Report by Dr Henry Littlejohn, Edinburgh's first medical officer of health,[21] the city was almost completely surrounded by irrigated meadows, jealously guarded against municipal bye-laws by landowners raking in high rents.

Sewage, however, turned out to be too dilute to be very effective except for vegetables and the cost of pipes was often not covered by increased yields. This did not stop many keen agriculturists from advocating it, especially for cottagers, although it was always controversial. Many towns built sewage farms in the hope of profiting from selling the produce, but the fashion eventually waned towards the end of the century in the face of both practical problems such as storage in winter and increasing health concerns.[22]

THE COTTAGE MIDDEN

To cottage middens, made up of all the household and human waste from the house, could be added weeds and road-scrapings collected by the children. But it was dung from the family cow, if they had one, pig and poultry that was most valuable. As potato cultivation increased, the pig became almost more cherished for its dung than for pork and bacon. Farm servants' gains nearly always included either straw, if a pig was kept, or manure: in the 1890s, two loads of manure,

---

18 Ibid., 52.

19 Nicholas Goddard, '19th century recycling: the Victorians and the agricultural utilisation of sewage', *History Today*, 31, June 1981, 32–6, 32.

20 Chadwick, *Sanitary Report*, 121.

21 Henry D. Littlejohn, *Report on the sanitary condition of the City of Edinburgh*, Edinburgh 1865, 89–90.

22 Goddard, '19th century recycling', 36.

valued at 6s., was still part of the wage, at least in the Lothians and the Central Belt, although not in the north or the Borders.[23]

Did this mean that cottagers, or at least farm servants, were better supplied with manure than farmers? It seems possible. Jeremy Burchardt, discussing English allotment-holders' manuring habits, estimates that the dung from one pig per allotment of a quarter of an acre 'was well beyond what contemporary farmers normally achieved'.[24] The smaller cottage gardens of Scotland may have enjoyed even heavier dressings of pig dung. Cottages (and their inhabitants too) were also regarded as an ideal source of liquid manure. No less an authority than Henry Stephen's *Book of the Farm* suggested in 1848 that cottagers with a cow could build a small tank to collect the dung – 'provided the cow were fed within doors in all seasons, with green and soft food',[25] which sounds unrealistic for the average cottar's cow; in any case, by this time cow-keeping by farm servants was dying out, although many cottagers probably kept them. The following year, HASS offered a gold medal for a report on 'the most efficient and economical mode of collecting and applying liquid manure adapted to cottages and farm-houses as well as to farm-yards and offices: cleanliness in the arrangement for cottages being attended to'.[26] Human manure, it was argued, was superior to guano, the miracle fertiliser being imported from Peru, as it produced better quality and heavier crops, as well as being much cheaper. John Wilson of Eastfield, Penicuik, was one of the entrants. His essay described his servants' cottages, which were in a single row 100 yards from the steading. They were one-roomed cottages with a 'good kaleyard' behind, surrounded by a well-kept hedge – in other words, typical of the overcrowded and comfortless rural homes of the time. The pigsty was also behind the houses and a pit three feet deep had been dug beside it to take both the dung from the sty and the household waste. 'This is daily saturated with the rich liquid from the houses, which, along with the urine from the pig-house, renders it a valuable mass.' Nothing was thrown out of the front of the house, all waste material being carried round to the midden at the back, as there was no back door – so it could not be cleaner (he claimed). Ten to twelve cartloads of muck per year were collected in this way and used to manure the potato grounds, and any excess was easily sold. The author pointed out that the addition of the 'rich liquid from a cottage' to dry waste material such as ashes and pig dung made muck much more valuable.[27]

As well as garden crops, pot plants were thought to benefit from applications of diluted sewage. An article in a gardening magazine of 1871 recommended

23 *RC Labour* 1893, vol. III, pt II, Selkirk and Roxburgh, 56, 61, 76; pt I, Moray, Nairn and Banff, 115, 124, 156.

24 Burchardt, *Allotment Movement*, 148.

25 Stephen, *Book of the Farm*, vol. II, 412.

26 'Abstract of reports on collecting and applying liquid manure', *THASS*, 1849, 253–75, 253.

27 Ibid., 272.

adding fifty or seventy gallons of water to a bushel (eight gallons) of 'house sewage' mixed with horse, cow, sheep and deer dung as well as soot (the colour of sherry when properly mixed), although such quantities and the problem of storage might have made the most competitive cottage gardener quail.[28] Outside privies were often constructed to drain into covered ashpits from which more 'valuable manure' could be excavated.[29]

As the sanitary movement slowly eroded the middens, reliance on animal dung increased and the question of adequate manure supplies became, if anything, more vital. While cottagers who were paid in cash and did not keep pigs, cows or poultry would have struggled to acquire enough manure for their gardens, it is perhaps a measure of the progress of mixed farming that by the end of the century farmers were quite often in a position to offer their hinds unlimited supplies of manure from the cattle court. By the early 1900s cheap artificial fertilisers had become available to cottage gardeners, although whether they bought them or continued to rely on animal dung or middens is not known. Gardening books and magazines, though, tended to refer to 'manure' without specifying the type.

THE CLEANSING OF THE COTTAGES

Middens right outside the front doors of cottars' dwellings were as common a sight in the Scottish countryside of the eighteenth century – and much later – as the stinking piles of excrement in the wynds and closes of the cities (Plate 20). Travellers were more struck (sometimes literally) by the latter, perhaps because the smell was worse and the contrast with English and continental cities greater: the enormous height of the Edinburgh tenements, the density of population and the cramped streets and alleyways made waste disposal exceptionally difficult for the city authorities and encouraged the notorious practice of emptying chamber pots from windows. In villages and the countryside, passers-by at least avoided this menace.

Long before science had understood, and authorities had applied the knowledge of, the causes of disease, the need to separate human dwellings from dung heaps had begun to be appreciated. George Robertson noted that before the 1760s cottar houses in Midlothian were frequently sited opposite the farm midden.[30] Improvement saw the gradual detachment of the farm midden, if not its cottage cousin, from the farmhouse and the cottages on several grounds, including health.

---

28 Anon., 'Liquid Manures', *Journal of Horticulture and Cottage Gardener*, 1871, 218.

29 *RC Labour*, vol. III, pt I, Aberdeenshire and Kincardineshire, 135.

30 Robertson, *Rural Recollections*, 76.

Plate 20 Dung Heap in Gauldry, Fife, 1893.

Cottages had a single door at the front, and the most convenient place for the dung hill was just outside that door.[31] Not only concern for the health of farm workers but also the refining of sensibilities – at least those of the passing gentry – began to make the sight and smell of dung hills along village streets or in front of farm cottages offensive. Gardens at the front and the middens at the back was seen as one solution, although not everywhere. Planned villages, from at least the 1730s, rarely featured front gardens. Feuing landowners instead specified the houses should front directly onto the street to avoid the front door midden. But on the Adam estate in Kinross cottages to be built for labourers and tradesmen in the 1790s were to have the garden in front, 'close before the door', with the dung hill and byre behind. William Adam, son of the architect, writing to Sir John Sinclair as president of the Board of Agriculture about the improvements he had already achieved, argued that 'this, if [the cottagers] have any turn at all for neatness, will give them an opportunity to exercise it; and if they have none, it will exclude the common nuisance of our Scotch cottages, the dung and filth in front.'[32] In spite of his 'trade' origins, Adam was one of Scotland's leading political managers at this time, corresponding with the aristocratic and

[31] Not only in the countryside: at the entrance to Crieff, the second-largest town in Perthshire, dunghills in front of the houses almost blocked the road (NRS E729/8, Archibald Menzies, 'Report on the Perth Estate', Annexed Estates papers, 93).

[32] Adam, 'Improvements on Blair', 8.

political elite, including royalty. It seems likely that he was perfectly in step with his enlightened peers in his ideas on cottage design.

The battle against the front-door midden was waged vigorously by the Highland and Agricultural Society of Scotland. One of the rules for their best-kept cottage garden competition was that dung hills be removed from the front and the side of houses, although the doubtful success of this competition may have limited its effectiveness. The problem remained, but the gradual elimination of the front-door midden was gratifying: 'the miniature flower-garden in front (instead of a midden and a puddle), the China roses against the walls, the back-door, covered ash-pit, etc, and the neat kitchen-garden' were described by Mr W. Henderson, 'a very intelligent nurseryman', responsible for improved cottages on a Scottish estate.[33] And in Berwickshire 'not only have dung-hills and all other deformities been generally removed, but in many cases the fronts and gables of the cottages have been ornamented with honeysuckles, roses, and ivy.'[34] But such pleasing sights seem to have been uncommon. 'The stagnant puddle and the nauseous dunghill' in front of the cottage persisted, even when the dwelling was neat and clean inside, and the inhabitants equally so.[35]

CHADWICK'S SANITARY SURVEY AND ITS AFTERMATH

Edwin Chadwick visited Edinburgh in the course of his work on the first national survey of public health and investigation into the connections between poverty, dirt and disease. While there he met HASS members and commended the best-kept cottage premiums, 'which have everywhere excited competition, and have been attended with beneficial results'.[36]

The general sanitary condition of rural workers in Scotland, however, was dire. The reports Chadwick received from doctors working amongst the labourers, colliers, fisherfolk and hinds of East and Mid Lothian paint a wretched picture and provide a salutary counterweight to the generally upbeat accounts in the *NSA* and contemporary newspaper reports. They make it abundantly clear that for the majority of rural and village cottage dwellers in the 1840s the midden remained the only place outside the home where refuse could be disposed of.

The report for the mining village of Tranent in East Lothian and its neighbouring villages was written by Dr Scott Alison, the local medical practitioner. His main purpose was to catalogue the high mortality and the diseases of the poor – typhoid was the biggest killer – and to describe the appalling conditions that caused them. Amongst many other details, he recounts how the poor

[33] *The Cottage Gardener*, 1849, 238.

[34] W. S. Gilly, *The Peasantry of the Border: an appeal on their behalf*, Berwick on Tweed 1842, appendix 1.

[35] *NSA* Kinglassie, Fife, vol. IX, 195.

[36] Chadwick, *Sanitary Report*, 333.

maintain stinking heaps of 'ashes, cinders, fulzie [filth, excrement], and animal and vegetable waste' before their doors in order to sell them to neighbouring farmers, for between 1s. 6d and 4s. the cartload. It was this trade in muck that kept the streets piled high with refuse, only cleaned when the farmer who had contracted with the proprietor for the purpose decided to collect it. Colliers kept their middens outside the front door too, but 'in some of these houses the females are so lazy, and filthy in their habits that they carry their ashes and cinders no further than to a corner of the apartment, where they accumulate and have their bulk swollen by the addition of various impurities', which might include dung from the resident poultry and horse. Houses were often damp from standing in a permanent stagnant puddle.[37] The hinds on the other hand mostly lived in recently built, comfortable houses, albeit with only one apartment and a floor of beaten earth, and kept their middens at the back – perhaps because, instead of selling the contents, they used them on their own well-cultivated gardens.[38] Day labourers, inhabiting older houses with leaking roofs, windows without glass and no ceilings, had, according to Dr Alison, worse habits and, like the colliers, they were often drunk.

The report on Musselburgh and Fisherrow, also in East Lothian, and also compiled by the local doctor, described how the colliers living on the edge of Musselburgh had small gardens in front or behind. Too often these patches of ground were used simply to collect mounds of refuse:

> in the village of Craighall, entirely inhabited by colliers, there is a square formed by the houses, the fronts of which look inwards, and the centre divided out into small plots for growing vegetables or the like – but in most instances the pieces allotted to each are either a barren waste or covered with all the filth and refuse that they could collect, which, I believe, after they have got a certain quantity, they sell for manure.[39]

The practice of selling cottage refuse was found in cities, towns and villages all over Scotland, in spite of attempts to police it, at least in the burghs. Even in St Andrews, one of the less dirty towns ('the South Street is even clean'), 'the filth is made to pay for its own removal',[40] while in the suburb of Argyle, where almost every house had a garden behind, 'it is a matter of no little importance to be the owner of a dunghill' that adorned the front of each house, with the pigsty beside it. These were the homes not of the poorest inhabitants but of relatively well-paid and respectable artisans and tradesmen.[41]

37 Ibid., 85.
38 Ibid., 208.
39 Ibid., 136.
40 Ibid., 271.
41 Ibid., 271.

Chadwick encountered an exception to this miserable picture on a visit to the cotton mills at Deanston in Stirlingshire. There he found that rubbish from the workers' houses was cleared away every morning and taken to a communal dung heap away from the village, from where the villagers could cart it to their gardens. Chadwick believed that these villagers were healthier than both the mill workers who lived outwith Deanston and the agricultural labourers.[42]

Progress in sanitation following Chadwick's revelations was painfully slow in Scotland. The Association for Promoting Improvement in the Dwellings and Domestic Condition of Agricultural Labourers in Scotland was formed in 1855 by progressive landowners and ministers. In its fourth *Annual Report*, Alexander Thomson of Banchory (Kincardineshire), a regular contributor to agricultural journals, related the snail's pace of improvement. While dung hills no longer stood before the cottage door, still 'it will generally be found that there is a certain spot, about a yard in circumference, and coming to within a foot or two of the door, on which the bustling goodwife deposits all her smaller slops without compunction.'[43]

Rural districts suffered the scourge of cholera, typhus and other diseases caused by polluted water and open sewers, although less than the cities. But the 1850s also saw the start of some improvement. *The Cottage Gardener* magazine, perhaps stretching the truth somewhat, compared the present scene to ten or twenty years before: 'miserable huts still exist but the cottages wear a greater air of cheerfulness and comfort; the dirty pool and midden (alias dunghill) close to the door and window have, in many cases, given way to a pretty flower plot'.[44]

Hiding the midden at the back of the cottage frequently merely shifted the problem, even while it soothed the sensibilities of passing onlookers. Jane Brown vividly describes her own (English) cottage with its backyard, like so many, 'trampled by tiny feet, repository of washing water and general rubbish, and of dead, dead soil [which] could not have been gardened … it was simply worn out of existence'.[45] In Scotland, where the backyard midden remained part of the cottager's income stream, this part of the back garden was probably in a worse state.

The midden maintained its popularity, both with farmers and cottagers. While guano was being used on Scottish farms from the mid-1840s, along with saltpetre and other artificial fertilisers, household refuse was cheaper and dung heaps represented a valuable commodity for cottagers – 'the chief treasures of the poorer cottagers and labourers'.[46] The town council or a proprietor might sell street refuse from an entire village to local farmers and

42 Ibid., 306.

43 *4th Annual Report of the Association for Promoting the Improvement of the Dwellings and Domestic Condition of Agricultural Labourers in Scotland*, Edinburgh 1858, 20.

44 *The Cottage Gardener*, vol. 6, 1851, 128.

45 Brown, *Pursuit of Paradise*, 150.

46 *Annual report by the County Sanitary Inspector Inverness-shire (Scotland)*, Inverness 1892, 304.

so employ a scavenger to – occasionally – clean them, but individual cottagers often guarded their heaps jealously, building them up for months until they had accumulated a saleable amount.[47]

An understanding of the health risks of spreading untreated sewage on fields and gardens was slow to arrive. The miasma theory of infection – that it was poisonous gas, containing particles of decaying matter, that caused disease – persisted until the 1880s, so diluting sewage with river water, it was held, was all that was needed to reduce if not eliminate its health risks.[48] As late as 1891, farmers in East Lothian, Perthshire and Angus were collecting the contents of cottagers' backyard ashpits to spread on their fields.[49] In Downfield, on the edge of Dundee, sewage was still irrigating the fields in the 1890s.[50] Municipal refuse known as 'town dung' or 'police manure' was transported by train or lorry from Dundee to the surrounding farms to be spread on the field until at least the 1930s. The fragments of metal, glass, china and clay pipes that were mixed in with the 'dung' are still being dug up by detectorists.[51]

While Chadwick's *Report* led, eventually, to the English and Welsh Public Health Act of 1848 and the assumption of responsibility for clean water and sanitation by local authorities, for Scotland (well behind England and Wales, as in much social reform legislation) a Public Health (Scotland) Act giving the poor law Board of Supervision powers over public health was not passed until 1867.[52] And, twenty-four years later, many rural areas still lacked drains and access to clean water. The Sanitary Inspector for Angus wrote in his 1891 *Report*:

> in the rural parts of the County, no provision has been made for drainage … except for the better-class homes or houses of recent construction; hence all slops or other foul water are usually deposited in ashpits or thrown into roadside ditches, or become nuisances. Where drains exist, they either discharge into streams or ditches, or cesspools, and the liquid overflowing therefrom generally finds its way to streams.[53]

Ashpits and privies may have made their first appearance in workers' dwellings built on the Rosebery estate near Edinburgh in the 1830s and inspired

---

47 NRS, HH62/2, Medical Officer of Health Reports 1891 Roxburghshire, 18.

48 Flinn, 'Introduction', in Edwin Chadwick, *Report on the Sanitary Condition of the Labouring Population of Great Britain* [1842], ed. Flinn, Edinburgh 1965, 59.

49 NRS, HH62/1, 33, East Lothian; HH62/45, 55, Perthshire; HH62/1, 15, Forfar (Angus).

50 NRS, HH62/1, 9, Forfar.

51 Personal communication from David Drummond, 9th August 2017.

52 There may have been several reasons for this tardiness, including suspicion of centralised bureaucracy, the persistence of the miasma theory and the more recent reform of the poor law and of civil registration of births, marriages and deaths (Flinn, 'Introduction', 72–3).

53 NRS, HH62/1, 9, Forfar.

by an influential article by an architect on cottage construction. Each cottage had an outdoor privy and an ashpit, both at the back.[54] Whether the ashpits were covered is not made clear, although later in the century they generally were.[55] Nevertheless, such conveniences were adopted country-wide extremely slowly. By the end of the century the majority of workers in central and south-eastern Scotland were living in cottages built in the previous thirty years. Official reports agreed that on the whole these cottages were drier, more comfortable and larger than before. However, the first reports of the new County Medical Officers of Health and Sanitary Inspectors (appointed under the terms of the 1889 Public Health (Scotland) Act) present a shocking picture of some workers' dwellings, fifty years after Chadwick's Inquiry. James P. Watt's account of Ellon District in Aberdeenshire can stand for many others: 'Damp roofs, damp walls, damp earthen floors, imperfect ventilation, the old unhealthy boxbed, badly constructed ashpits and privies, dung heaps close to windows or doors, pumps or shallow wells in close proximity to ashpit, midden or midden drain'.[56] Many of the newer houses had been built into hillsides and damp was still endemic.

Villages and cottages were still blighted by middens. In Roxburghshire 'the absence of ashpits leads to the formation of middens unpleasantly close to houses.'[57] Where there were ashpits, they were nearly all uncovered, polluting the surrounding soil. Scavenging in villages was almost non-existent. In a village near Kelso there was only one ashpit, many houses had no drains and there had been no street cleaning since the scavenger had been paid off. There were several 'large and offensive middens'.[58] The Medical Officer for Fife and Clackmannan noted liquid manure leaking from pigsties and contaminating drinking water in badly sited and badly built rows of cottages.[59] In Aberdeenshire Watt concluded that the prime need was for scavenging. Cleaning the streets would mean the dung could be stored in central depots from where farmers could buy it – at a lower price, as costs would be reduced.[60] Even where there were sewers, as in several villages in Angus, they were neither efficient nor ventilated.[61] Everywhere,

[54] According to the author, the privies had to be outdoors as there were no plumbers in the countryside and labourers preferred outdoor privies to indoor water closets. George Smith, 'Essay on the construction of cottages suited for the dwellings of the labouring classes, and adapted to the climate of Scotland', *THASS*, 1835, 205–16, 208.

[55] After the 1850s and up to the first decade of the twentieth century, a period when interest in and the fashion for cottage building declined, there is no mention of ashpits or privies in *THASS*.

[56] NRS, HH62/1, 20, Aberdeenshire.

[57] NRS, HH62/2/17, Roxburghshire.

[58] Ibid., 27.

[59] *RC Labour*, vol. III, pt II, Fife, Kinross and Clackmannan, 71.

[60] NRS, HH62/1/17, Aberdeenshire.

[61] NRS, HH62/1/15, Forfar.

unemptied ashpits (often uncovered), polluted burns, contaminated wells, very few privies and inadequate or broken drains were to be found.

The medical officers and sanitary inspectors, whose duties were endless,[62] could recommend improvements, but it was up to the county councils to decide whether or not to implement them. The Public Health Act (Scotland) of 1867 and a subsequent Act in 1889 did not give local authorities the powers to compel scavenging by prosecution. By the 1897 Act[63] county councils could still not prosecute but could exercise their powers to remedy nuisances by, for example, building or enforcing the building of drains and sewers and providing clean water.[64] But implementation of this Act was patchy and the age of the midden as the destination of all the waste produced by a cottage and its inhabitants did not finally come to an end until well into the twentieth century.

In such conditions of filth and degradation, which seem for so many to have endured for so long, any attempt at gardening will seem to modern eyes miraculous.

62 Duties ranged from the regular inspection of dwellings, villages and food shops to the monitoring of drinking water and fumigation of infected houses (NRS, HH62/2/5, Roxburghshire).

63 Public Health (Scotland) Act 1897. Available at https://www.legislation.gov.uk/ukpga/Vict/60-61/38/contents/enacted.

64 I am grateful to Dr Sarah Hendry for explaining the law on this.

CHAPTER 6

THE RURAL DIET

Thus hae we played through a long simmer's day,
We thocht on nae ill, and we dreaded no wae;
Till ance the bricht sun fell a-hiding his nose,
And then we gaed hame to get our kail-brose.
A dish o' guid brose wi' the kail and the says
A herdie was needin' jist aff the leas; [meadows]
The kail and the says and the drappie o' ream [cream]
Wad set me a-sleepin' as seen's they were deen.'

The Herdie[1]

> Most eminent physicians of the last age were in use to ascribe a remarkable improvement in the health of the Scottish peasantry, – the alleviation of some prevailing diseases, and the extirpation of others – to the general introduction of cottage gardens and consequent increased use of coleworts [kale], cabbage and potatoes.[2]

Thus Patrick Neill in 1813. Not only was this diet more wholesome than before, it was cheaper and so appealed to the Scots' sober and frugal habits. It was also, as he noted, based on the garden: the ingredients of Scotch Broth – 'the national dish' – 'are chiefly derived from these gardens'. These ingredients include leek, onion, cabbage or kale, turnip and carrot as well as pearl barley and, for those who could afford it, a piece of mutton or beef. Scotch or barley broth is still popular today.

However, the aversion of modern Scots to vegetables of any kind other than in broth and with the exception of potatoes is notorious. Recent studies document the significantly lower consumption of vegetables and fruit in Scotland today, compared to England.[3] All social classes appear to share the habit of vegetable-

1 Ord, *Ord's Bothy Songs*, 270.

2 Neill, *Gardens and Orchards*, 18.

3 Paul Scarborough et al. calculated that consumption of fruit and vegetables in Scotland was 20% lower than in England (Scarborough et al., 'Differences in coronary heart disease, stroke and cancer mortality rates between England, Wales, Scotland

avoidance, but the poorer you are the less likely you are to eat five portions of fruit and vegetables a day (and the less likely you are to be able to afford them).[4]

Was Neill correct about the importance of vegetables in the rural diet in the early 1800s, and if so, what happened to it later? This chapter investigates the food and the meals of rural workers and their nutritional value, as they changed during the period.

'Even at a nobleman's table we had only flesh meat of one kind, and no vegetables of any kind', recalled John Wesley after his first visit to Scotland in 1762.[5] Wesley may have just been unlucky, as walled gardens equipped with elaborate hothouses had been raising the most exotic and varied fruit and vegetables for the tables of the Scottish aristocracy from at least the end of the 1600s. The first Scottish gardening book, *The Scots Gard'ner*, published in 1683, listed over thirty vegetables suitable for the kitchen garden.[6] On the other hand, it took some time for such a wide range of vegetables to reach even middle-class tables. In the 1740s and 1750s, when Thomas Somerville was growing up in the manse in Hawick in the Borders '"open kail", cabbages, turnips[7] and carrots were the only vegetables in general use'.[8] By Wesley's second visit in 1780 vegetables were as common on upper-class dining tables as in England.[9] Vegetables were being grown commercially by market gardeners and sold in city shops and markets. 'The demand for vegetables [in Edinburgh] has increased ten-fold within these fifty years', it was claimed.[10]

The recipe book published by 'Meg Dods' in 1826[11] noted that

> we now see a regular succession of not merely broccoli, cauliflower and pease, but of the more recondite asparagus, seakale, endive and artichoke, with an abundance of small saladings … . The vegetable markets of most towns have

---

and Northern Ireland: the role of diet and nutrition', BMJ Open 2011;1:e000263. Doi: 10.1136/bmjopen-2011–000263, 3).

4 David Marshall, Annie S. Anderson, Mike Lean and Ann Foster, 'Healthy eating: fruit and vegetables in Scotland', *British Food Journal* 1st August 1994, 18–24; Scottish government, *Scottish Health Survey 2012*, 2013, available at https://www.gov.scot/publications/scottish-health-survey-2012-volume-1-main-report/pages/16/.

5 Quoted in Cox, *Gardening in Scotland*, 118.

6 John Reid, *The Scots Gard'ner*, Edinburgh 1683.

7 Swedes, known as turnips (neeps) in Scotland, were introduced only in the 1750s, while white turnip was served at upper-class tables (Alexander Fenton, *The Food of the Scots*, vol. 5 of *Scottish Life and Society, A Compendium of Scottish Ethnology*, Edinburgh 2007, 208).

8 Thomas Somerville, *My Own Life and Times 1741–1814*, Edinburgh 1861, 330.

9 Cox, *Gardening in Scotland*, 118.

10 *OSA* Inveresk, East Lothian, vol. XVI, 16.

11 'Meg Dods' [Isobel Christian Johnston], *The cook and housewife's manual: a practical system of modern domestic cookery and family management*, Edinburgh 1829.

> within the same period undergone a wonderful improvement ... so that a healthful luxury is now within the reach of all classes.[12]

Although this was an extravagant claim, the importance of fruit and vegetables for health was beginning to be recognised. Fruit and vegetables had been traditionally regarded in much of northern Europe with some suspicion as the food of the poor who could not afford meat or cereals, or even as causing fevers.[13] Medical opinion, meanwhile, came to appreciate their health benefits: a healthy body needed a balance of acidity (from fruit and vegetables) and alkilinity (from meat). Some were also beginning to understand that vegetables could prevent scurvy, particularly after the experience of the potato famine of the 1840s.

Many nineteenth-century nutritionists, led astray first by Liebig's insistence on energy deriving only from protein and then by the discovery of germs, abandoned theories that explained scurvy and other conditions as deficiency diseases.[14] So the surveys of the diets of rural workers of the 1860s, for example, described later in this chapter, and long before the identification of vitamins in the early twentieth century, calculated protein and carbohydrate intakes and mostly ignored vegetables and fruit. These were considered useful only for their acidic salts and for variety, and were to be avoided by nursing mothers and eaten in moderation by everyone else.[15]

The diet of the working class and in particular of agricultural workers was of considerable interest and concern in the eighteenth and nineteenth centuries. It was recognised that the continuing progress of agriculture depended on healthy and vigorous ploughmen and labourers. Most of the information we have today on the diet therefore relates to agricultural workers, and especially to farm servants. In spite of the differences in geography and social and economic organisation across the Lowlands, that diet was remarkably uniform, dominated throughout the period not in fact by vegetables or even potatoes but by oatmeal. In the north-east oatmeal remained a central part of crofter life well into the second half of the twentieth century. Alexander Fenton, recalling life on an Aberdeenshire croft in the 1960s, recounts that oatcakes, known as 'breid', were

---

12 Quoted in F. Marian McNeill, *The Scots Kitchen*, new edn, Edinburgh 2010, 18.

13 There is a single, intriguing, comment in the *GV* for East Lothian: 'formerly the common people were fed mostly upon vegetables'; now, their food was improved, with porridge for breakfast and supper and peasemeal and barley bread. *GV East Lothian* 1794, 104. J. C. Drummond and Anne Wilbraham, *The Englishman's Food: a history of five centuries of English diet*, London 1957, 234; A. Kowalska-Lewicka, 'The Potato and the Polish kitchen', in A. Fenton and Eszter Kisban (eds), *Food in Change*, Edinburgh 1986, 90–5, 92.

14 Kenneth J. Carpenter, 'A short history of nutritional science: Part 1 (1785–1885)', *Journal of Nutrition*, vol. 133, 2003, 638–45, 641.

15 F. W. Pavy, *A treatise on food and dietetics, physiologically and therapeutically considered*, London 1874, 119.

still the accompaniment to every meal,[16] although wheaten bread ('loaf bread') and scones and pancakes made with wheaten flour had become everyday fare. An old Aberdeenshire labourer remembered at the end of last century:

> Oh there wis a lot o oatmeal used. Brose, at wis the main breakfast, aye, as ye ken [know]. And oatcakes ye see, wis used an affa lot, and skirlie [oatmeal mixed with chopped onions and fried in suet or fat] and the porridge, oh no oatmeal wis the staple diet, and tatties an affa lot. But good oatmeal and good tatties, oh I suppose it got monotonous, but if it wis cooked richt and prepared richt there wisnae much adee [the matter] wi it.[17]

Oatmeal graced special occasions. At the birth of a child sugared oatmeal bannocks were given to the midwife and the other women who came in to assist. At weddings the bridal cake – often a sweet oatcake – was broken over the bride's head. A dish of fermented oat husks called sowens made the Christmas Eve, or Sowens Nicht, meal.[18] According such an ordinary food such high status may seem strange, but oatmeal was never the food of the very poor, and is not even an age-old tradition. Although it formed part of the diet of some of the ordinary people before the late 1700s, oatmeal was much too valuable to be the standard fare of the poor. Oats were not then a hardy crop and were quite likely to fail in bad harvests. They were more often grown as a cash crop and tenant farmers also paid their rent, servants' wages and mill dues all or partly in oatmeal.[19]

Medieval Scotland had enjoyed a diet in which meat and dairy produce predominated and cereals and meal from beans and peas started to take over only from the sixteenth century.[20] Gibson and Smout characterise this change from a diet based on animal products to one made up overwhelmingly of cereals as a deterioration.[21] Enid Gauldie's work on the Scottish country mills has shown that watermills were set up mainly to grind peas, beans and bere, or bear (a primitive form of barley), not oats. These different types of meal were often grown together and so milled together before being baked into flat bread (mashlum or masloch), boiled for porridge, or used to thicken kale broth. In Auchterarder in Perthshire in the 1770s 'a few tenants excepted, no family had oatmeal in their houses, nor could they get any. They had nothing better than bear meal and a few greens boiled together at mid-day and bear-meal porridge

16 Fenton, *Food of the Scots*, 6.

17 Anonymous Aberdeenshire witness, School of Scottish Studies oral history archive, 1994.

18 Fenton, *Food of the Scots*, 141, 146.

19 Enid Gauldie, *The Scottish Country Miller 1700–1900*, Edinburgh 1981, 2–3.

20 A. J. S. Gibson and T. C. Smout, *Prices, Food and Wages in Scotland 1550–1780*, Cambridge 1995, 226.

21 Gibson and Smout, 'Scottish food and Scottish history 1500–1800', in R. A. Houston and I. D. Whyte (eds), *Scottish Society 1500–1800*, Cambridge 1989.

evening and morning.'[22] Bere- or barleymeal and peasemeal remained cheaper than oatmeal, and beanmeal was even cheaper. The labourer poet of the 1830s, Alexander Bethune, complained:

> And the coarse bannocks baken from the bean
> (which last I never liked, though compelled
> To take them for my dinner many a day) ... [23]

Meanwhile, peasemeal and beanmeal remained a staple until the end of the nineteenth century, by which time it had been relegated to the animals, except for pease pudding and pea soup.[24]

It was only with the advent of improved agricultural practice and more reliable varieties, developed after some disastrous harvests, that oats could be grown to the extent that it became part of the wage and hence the daily food of the rural workers.[25] Enormous amounts were consumed: the six and a half bolls that made up part of the farm servant's wage translates into 36oz per day for the family, although some of the allowance would be sold to cottagers and others. The two great advantages of oatmeal were ease of cooking and variety of uses. With salt, water or milk, cream, butter, sugar, even ale or whisky, oatmeal could be boiled in a cauldron and made into porridge, pottage, gruel, brose, crowdie or sowens. Vegetables added variety: kail brose was made by cooking the meal in the water the kale had been boiled in, and neep and even nettle brose in the same way. Thus vegetables became integrated into the cereal diet in a way that was not possible in the bread-based diet of England. Oatmeal and the other meals could also be baked into cakes or bannocks. Mixed with fat or suet, and onion or animal blood, oatmeal became a pudding to be boiled, and, with sheep's offal, haggis. In all these different guises, it was enjoyed by rich and poor, and from north to south. Different parts of the country had decided preferences: in the north and north-east fine oatmeal was preferred; in the Central Belt, they liked a coarser milling, or even pinhead.

Meal was delivered from the mill two, three or four times a year, the oats being stored in the girnal or kist in the kitchen. Alexander Somerville's labourer father bought enough bolls of oats to feed the family for a year. These were then ground into meal by the miller, with the help of Alexander's mother for the sifting:

> During the afternoon, the 'lademan' brought home the sacks of meal piled upon his double-horsed cart, with probably my mother sitting on the top of the sacks. The meal was carried into the house by the man; he got a dram of

22 Quoted in Gauldie, *The Scottish Country Miller*, 3.

23 Bethune, *Memoirs*, 25.

24 Fenton, *Food of the Scots*, 201.

25 Ibid., 8–9.

> whisky according to custom, and drove homeward, cracking his whip with an air of importance peculiar to him and all millers' lademen.[26]

At night the family had new mealbrose for supper, which was a high treat – 'the old oatmeal of the year before having become damp and bitter ... the new meal, though made into 'bare brose', was so agreeable when it first came home that to put anything richer than itself to it [such as milk or butter] was deemed a waste.'[27]

Farmers grumbled later in the century – when labour had become hard to keep on the farm – that their servants would accept only the very best quality, sometimes obliging them to buy in the meal when the home-grown crop was not deemed satisfactory.[28] Ploughmen who gave evidence to the Royal Commission on Labour in 1893 had no complaints about the 'Ploughmen's meal' because it was the best,[29] although others did complain.

COOKING

It was the cooking arrangements in the cottages that determined the dishes that could and could not be made. Until quite late in the nineteenth century ovens for baking and even hobs were in very short supply. Designs for farm cottages from the 1830s include fireplaces in the gable ends and grates were in use from at least the 1820s, when Alexander Somerville's family had them, and took them with them at every flitting.[30] But in most country homes in the first half of our period fireplaces were marked out on the beaten earth floor in the middle of the main room (the but) with a circle or square of flat stones, as they had been for centuries.[31] The smoke, from a peat, turf or brushwood fire, escaped (if the wind was in the right direction) through a hole in the roof. Cooking was done in a pot (kailpot) suspended by an iron chain or a grass or heather rope either from the roof ridge or a crossbeam. For porridge, coarse oatmeal was soaked overnight then cooked in the pot for around forty minutes; finer types took between ten and fifteen. Brose was made in just a few minutes by pouring a handful of oatmeal into boiling water. Oatcakes and bannocks were cooked on an iron girdle (or griddle) also suspended above the fire.

---

26 A. Somerville, *Autobiography*, 44.

27 Ibid., 44–5.

28 Shirra Gibb, *A Farmer's 50 Years*, 17.

29 *RC Labour*, 1893, vol. III, pt 1, 160.

30 Alexander Fenton and Bruce Walker, *The Rural Architecture of Scotland*, Edinburgh 1981, 144.

31 Until the eighteenth century, the inhabitants sat and slept around this hearth on earthen beds (Fenton, *The Hearth in Scotland*, Scottish Vernacular Buildings Working Group, Edinburgh 1981, 9).

Plate 21 Miss Lizzie Robertson by the hanging lum, Moirlanich, Perthshire, c. 1930.

Later a free-standing stone slab was placed behind the fire, and under it a deep ashpit, emptied from time to time, the ashes used for animal bedding or manure for the garden. Hearths were moved to the gable end, where there was now a chimney, a wooden or lath-and-plaster 'hingin' lum', attached to the gable (Plate 21). A fireplace built against this wall could be fitted with a swee, from which pots and kettles and girdles hung.[32] So long as peat, wood or turf was the fuel hearths could be at floor level, but coal – increasingly common as the wasteland where peat and brushwood could be sourced gradually disappeared under cultivation – needed a raised fireplace for the under draught.[33] The stronger heat from a coal fire also required a chimney built into the gable end and wrought iron grates and hobs; cast-iron ovens came later. Robert Hutchison was recommending 'fixed in' grates with a boiler

[32] Ibid., 17–32.

[33] Hugh Cheape and Gavin Sprott, *Angus Country Life*, Edinburgh 1980, 24.

for cooking (and an extra boiler for the pig's food) in the late 1860s, suggesting these were not yet standard in cottages.[34]

For most of the nineteenth century it was the kailpot or cauldron hung over the fire that cooked most of the meals. Soups, broths, porridge, brose, boiled meat, puddings, haggis and dumplings were all boiled, simmered or steamed in the pot. Boiling, simmering or steaming remained standard Scottish cooking methods into the twentieth century.

VEGETABLE DISHES

Rachel Laudan, making the point that 'most societies ... came to depend on two or three favourite roots or grains to provide most of their calories, that is, as their staples', notes that 'leaves and shoots [were] low in calories, often bitter, and difficult to store [and] were used mainly as medicine.'[35] So why eat them? What Sidney W. Mintz has termed 'peripheral' foods – everything that is not a grain or a root – may have initially made carbohydrates easier to swallow, and less boring, as well as adding essential nutrients people would have been unaware of.[36] In Scotland the peripheral foods known as 'potherbs' – mostly brassicas such as kale and cabbage, but also leeks and onions – were by the nineteenth century daily fare almost as essential as meal, for those with gardens at least, and far more common than meat or fish.[37] Those without gardens, notably the single farm servants in the bothies, often lived almost entirely on oatmeal, with the occasional salt herring or small piece of beef.

The commonest of the potherbs was kale. Kale has featured in every account of the Scottish diet for hundreds of years. Packed with vitamins and minerals, capable of withstanding the harshest winter weather and harvested throughout the autumn, winter and into spring, it is not surprising that it came to represent the diet perhaps even more iconically than oatmeal. It was so much part of the diet that the word 'kail' could stand for 'dinner'.[38] As late as the mid-twentieth century 'kail broth' or simply 'kail' was a common term for any broth. Nettle kail, for example, was made from foraged nettles. And the word appears in any number of proverbs, sayings and customs.[39]

---

34 Hutchison, 'Dietaries of Scotch agricultural labourers', 15.

35 Rachel Laudan, *Cuisine and Empire; cooking in world history*, Berkeley 2014, 1.

36 Sidney W. Mintz, *Sweetness and Power: the place of sugar in modern history*, New York 1985, 11.

37 Potherbs of all kinds did not receive their true nutritional credit until the twentieth century. Edward Smith described leafy vegetables as the 'least nutritious class of vegetable foods' (Edward Smith, *Foods*, 2nd edn, London 1873, 207).

38 McNeill, *The Scots Kitchen*, 27.

39 The *Dictionary of the Scots Language* website has many of these listed under 'kail' (http://www.dsl.ac.uk/entry/snd/kail_n).

Was kale eaten everywhere? Frequent references to scurvy in the *OSA* suggest that it was not, in spite of the fact that only a small amount of vitamin C (2–3mg) each day, which could be supplied by one ounce of kale, is enough to ward off the disease.[40] Gibson and Smout conclude that 'This must mean that in such circumstances, to all intents and purposes, no greenstuffs were eaten',[41] not even wild herbs at some times of the year. Kale was only ever an everyday food and, unlike oatmeal, never seems to have appeared in ceremonial or celebratory dishes. For most, it was the main 'greens' through three seasons of the year, until cabbage was ready in the summer.

The Mearns poet Joseph Grant wrote of the Deeside kail – the dish, not the plant – as much superior to the southern variety:

> The Deeside kail is not that heterogeneous mass of all manner of vegetables which our brethren south of the Grampians similarly designate – the kail in this district is composed of greens or cabbages alone, mashed, after boiling, with cream or butter, and, in this state, forms a really agreeable dish, whether hot or cold.[42]

Grant's account is outdone by another Aberdeenshire man. When William Milne arrived at a farm to take up his position as a ploughman (where, as a single man, he would be housed in the chaumer and fed in the farm kitchen), he recalls that

> I had forgot up till now that I was now in the home of 'seven kinds of brose', and the first dish put before us now was the celebrated 'curly kail brose', which, when well made is a dish fit for a king! This was followed by the greens mashed by themselves after being from the liquid, which, with the oatmeal and salt had formed the first dish. These mashed, properly boiled and mixed with cream, form a food compound or salad ... which by itself is really delicious, and tastes the nearest like prime boiled ox beef of anything I can compare it with.[43]

Said to have been brought to Scotland in the 1650s by Cromwell's troops – so a relative novelty – cabbage was the alternative to kale throughout the summer. Turnip (swede) became commonplace in the late eighteenth century. Introduced as an animal feed by the lowland improvers, it soon spread to the gardens of the poor. Neep brose often appeared as an alternative to kale brose in winter.[44] Onions, leeks and carrots were also added to the various kinds of broth and brose.

---

40 Gibson and Smout, *Prices, Food and Wages*, 242.

41 Ibid.

42 Joseph Grant, 'The warlock of Glendye', in *Tales of the Glens, with ballads and songs*, Stonehaven 1835, 80.

43 William J. Milne, *Reminiscences of an old boy: being autobiographic sketches of Scottish rural life from 1832 to 1856*, Forfar 1901, 193.

44 Fenton, *Food of the Scots*, 208.

In F. Marian McNeill's *The Scots Kitchen*, first published in 1929, the traditional vegetable dishes are nearly all made of boiled root vegetables, usually potatoes, mashed with milk and butter and mixed with kale or cabbage. Even Scottish recipe books in print today contain very few recipes for vegetable dishes, although many for soup and broth. Theodora Fitzgibbon's *A Taste of Scotland* (1970), including aristocratic as well as more humble recipes and several broths and stews – Howtowdie, Hotch Potch (or Hairst Bree), Stoved Chicken – that combine meat with vegetables, has no vegetable-only dishes. Jane Grigson's *English Food* (1974), by comparison, has nearly forty vegetable recipes, quite apart from soups.[45]

Potatoes very quickly took centre stage in the meals of all classes and became, for the poor, more essential than oatmeal. By the 1790s, according to the *OSA*, although potatoes were not yet in use everywhere, in most districts they had already become the staple food of the autumn, winter and early spring, and in some places meal is not mentioned at all: in Innerleithen, Peeblesshire, cottagers' 'wants are few and easily satisfied with milk and potatoes'.[46] Potatoes and meal (oatmeal and pease and barley meal) were standard items in the farm servants' wages by the time of the *NSA*, and are referred to as forming part of two meals a day: 'the ordinary food of the peasantry at their meals is oatmeal, potatoes, fish, and in most families, in the country part of the parish, a little pork of their own rearing.'[47] But there were some initial pockets of resistance. The taint of poverty attaching to this food meant that servants in Bendochy 'are commonly above eating potatoes' and in Alford in Aberdeenshire 'the common people are not very fond of them, and they think them unwholesome, nor will farm servants make a meal of them, or even eat them without milk or butter, so readily as in other parts of Scotland', although everyone grew a few.[48] Potato eating was less common in the north-east. In Morayshire potatoes did not appear in the labourers' diet at all,[49] although next door in Nairn potatoes and herrings were said to play a significant part in the daily fare.[50]

One of the many great virtues of potatoes was that, unlike oatmeal, which really needed milk or greens to be palatable, they could be eaten on their own with no added 'relish'. Archibald Gorrie wrote in 1832 that there were 101 different ways of preparing milk and potatoes.[51] Clearly there was something about potatoes that encouraged magical thinking. Milk (in the north-east) and herrings were nevertheless a favourite addition to potatoes in the latter half of

45 Theodora Fitzgibbon, *A Taste of Scotland*, London 1970; Jane Grigson, *English Food*, rev. edn, London 1992.

46 *OSA* Innerleithen, East Lothian, vol. XIX, 597.

47 *NSA* Abercrombie, Fife, vol. IX, 345.

48 *OSA* Bendochy, Perthshire, vol. XIX, 351; *OSA* Alford, Aberdeenshire, vol. XV, 418.

49 James Donaldson, *GV Elgin or Moray*, Edinburgh 1794, 24.

50 Donaldson, *GV Nairn*, Edinburgh 1794, 16.

51 A. Gorrie, 'Essay on Cottage Gardens', 72.

the century. Further south, potatoes might also be served with cheese, bacon and pork.[52] And, in spite of the deterioration in the rural diet towards the end of the century, boiled potatoes or potato broth for dinner and on Sundays beef and potatoes still ruled supreme.[53]

Rural families ate potatoes mashed or in broth, or plain boiled. Regional variation in potato recipes was limited. 'Meg Dods' has recipes for potato cakes, eaten with herring in the Highlands and Ireland, and 'plain Calecannon', a cottager's version of the Irish dish consisting of potatoes boiled with greens and mashed together with butter and a shredded onion.[54] F. Marian McNeill includes only four Lowland potato recipes. Two of the dishes are almost identical and derive from Calecannon: rumbledethumps came from the Borders and was made with equal quantities of cabbage or greens and potato mashed together, with butter. Kailkenny, from the north-east, was exactly the same, with cream instead of butter.[55] Stovies, still popular today and made by stewing potatoes and onions together with a little meat,[56] and potato scones completed the list.

Most of these vegetable dishes were prepared or taken with milk or milk products (although cheese was relatively uncommon), as were oatmeal dishes, and milk was the main drink. As we have seen, milk or a cow (and a cow's grass) was a highly prized element of the hind's wage. In the eighteenth century milk was usually skimmed or semi-skimmed, so most of the vitamins A and C and one-third of the calories were lost, even though the skimming process was less efficient than today.[57] Later, as turnips and sown grasses allowed better feeding of cattle throughout the year, more whole, or 'sweet', milk was available for drinking and turning into butter and cheese. As part of the farm servant's wage, for the increasing number of farm servants without a cow, one Scots pint (equivalent to three imperial pints) per day was the usual ration in the 1830s. Milk consumption gradually declined: the ration was reduced and when farms near large towns could easily sell their milk the rural poor often had to make do with treacle or sugar and water.[58]

---

52 Hutchison, 'Dietaries of Scotch agricultural labourers', appendix A.

53 Carter, *Farm Life in North-East Scotland*, 136.

54 'Meg Dods', *The Cook and Housewife's Manual*, 223.

55 McNeill, *The Scots Kitchen*, 180, 181.

56 *The Scots Kitchen* gives two places of origin for stovies – the Highland and the Maidens' Hospitals, forerunner of the Mary Erskine School in Edinburgh.

57 Gibson and Smout, *Prices, Food and Wages*, 237.

58 Hutchison, 'Dietaries of Scotch agricultural labourers', 5. Beer was the alternative to milk – it could be taken in porridge for breakfast – but beer-drinking seems to have been in decline throughout the period, thanks in part to high malt duties. Hutchison's 1869 report stated that beer was unknown as 'an item of the general diet' amongst farm workers in Scotland. Many commentators, notably in the two Statistical Accounts, expressed their unhappiness over these trends, and the substitution of whisky and tea for beer: 'the drinking of whisky instead of good ale is a miserable

From the late eighteenth to the twentieth century this very limited diet of oatmeal, potatoes, vegetable broth and milk fed the rural population morning, noon and night, leavened with salt herring, butter and cheese, and the very occasional piece of meat. Meat had been more common earlier in the century, but, according to some *OSA* reports, now appeared only at baptisms, marriages, holidays or Christmas. An apparent exception was the Carse of Gowrie, where, according to the *GV* published in 1794, 'Pork and bacon constitute a greater proportion of the food of the poorer inhabitants here than in any other part of Scotland, and there is scarcely a manufacturer or tradesman, or labourer, who does not feed one or two pigs every year, for the use of his family.'[59] Until the latter half of the nineteenth century meat of any kind barely figured in most workers' meals, even on a Sunday, apart from offal in haggis and mealy puddings and, as we have seen, in barley broth.[60] Once every cottager was growing potatoes and most were rearing pigs, bacon and pork[61] became commonplace, although in small quantities. Barrels of salted mutton too became familiar sights outside cottage doors. Only those living by the sea would have been able to add fresh fish and seafood to their meals.

What became of Scotch broth? There are remarkably few references to broth with or without meat in the *OSA* and only a few more in the *NSA*. It seems to have been a dish for Sundays or special occasions. Only in Newbattle near Edinburgh was broth 'and a little meat and potatoes and cheese' part of the 'ordinary food of the peasantry'.[62] It is likely that it was only when people could afford meat more often that it entered the standard diet. Hutchison's 1869 survey, however, found that broth – sometimes 'pork broth', sometimes 'vegetable broth', sometimes just 'broth' with or without bacon but always including vegetables – was consumed the length and breadth of the country and by almost everyone.[63] At the very end of the century, when diets were changing rapidly, Scotch broth, now with mutton, was still, according to R.

change, and so likewise is the very general use of tea. These put together have been exceedingly hurtful both to health and morals', wrote a north-east minister (OSA Mortlach, Banffshire, vol. XVII, 438). Whisky sealed every transaction, from the feeing fair to the delivery of meal from the mill. Tea-drinking, especially by women, was on the rise.

59 Donaldson, *GV Carse of Gowrie*, 30.

60 Maisie Steven, *The Good Scots Diet*, Glendaruel 2003, 66. 'One of the laddies was sent to the butcher with threepence for a sheep's heid and a penny to the blacksmith to singe the head at the forge' (Cheape and Sprott, *Angus Country Life*, 51).

61 Pigs had been raised in Scotland from Neolithic times. The spread of potato cultivation not only allowed cottagers to keep pigs but enabled improved breeding (Catherine Smith, 'A grumphie in the sty: an archaeological view of pigs in Scotland, from their earliest domestication to the agricultural revolution', *Proceedings of the Society of Antiquaries Scotland*, vol. 130, 2002, 718).

62 *NSA* Newbattle, Midlothian, vol. I, 70.

63 Hutchison, 'Dietaries of Scotch agricultural labourers', Appendix A.

Hunter Pringle, an East Lothian favourite, but it 'requires the addition of some green leaf to be true to character'.[64] The green leaf – cabbage and kale – still came from the garden.

VARIETY

To modern tastes this diet seems almost unbearably monotonous, but it should not be imagined that nothing else was ever eaten. Cottage gardeners raised fruit, especially berries, although fruit is almost never mentioned in contemporary accounts of workers' diets, and wild fruit and leaves were abundant in the countryside, as were, for some, poachable fish and game. Where there are references to fruit, it is clear that it was seen as a treat. Alexander Bethune's poem 'Day Dreams' (1834) tells us:

> Of gooseberries I reckoned not a little
> For I had bought a pennyworth and ate them,
> And oh! How sweet! I never could forget it'[65]

Bowls of the raw fruit were presented as treats to small children at fetes and newspaper reports of gooseberry thefts were not unusual. Were gooseberries sweeter in those days, or were teeth simply less sweet? Perhaps the latter: before the advent of cheap sugar (and the explosion of sweeties and rotting teeth), Scots seem to have preferred tart fruit; the old variety of apple the Bloody Ploughman, for example, was a dessert apple in Scotland but only used in cooking in England. But fruit was mainly cooked as it was considered indigestible when eaten raw.[66]

Wild berries and nuts were gathered, especially by children, but contemporary accounts rarely mention foraging, except in the Hebrides, where such exotic fare as seabirds and wild roots attracted a mixture of horror and fascination. In Perthshire watercress, sloes, hips, haws, wild raspberries, crowberries and whortleberries (from moorland), hazelnuts and crab apples were gathered. The poor infused juniper berries to make a tea.[67] Nettles, wild garlic and sorrel were also collected as food.[68]

In 1777 John Lightfoot, an English clergyman, published his two-volume *Flora Scotica* following a tour of the country. The book lists 1,300 plants, including twenty-four eaten as greens or salads in the spring. Lightfoot was intrigued by the 'otherness' of the Highlands and its inhabitants and gives much more

---

64 *RC Labour*, vol. III, pt II, Edinburgh and Haddingtonshire, 113.

65 Bethune, *Memoirs*, 21.

66 Fruit had too much water and too little nitrogenous content to be of value, and raw apples were difficult to digest, according to the nutritionist F. W. Pavy (*Treatise on Food and Dietetics*, 273).

67 *OSA* Clunie, Perthshire, vol. IX, 230.

68 Steven, *The Good Scots Diet*, 14.

information about the use of Highland edibles, while Lowland plants were often similar to those found in England. Lowland species such as the young leaves of nettles, Wild Spinage (sic) or Allgood, often eaten as greens (i.e. boiled), were 'very good tasted'.[69] Silverweed or moor-grass roots, which taste like parsnips, 'are frequently eaten by the common people in Scotland, either roast or boiled'.[70] These leaves would not, however, have protected against deficiency diseases such as scurvy in the absence of sufficient cultivated vegetables, although Lightfoot does mention at least one anti-scorbutic plant, pepperwort, which grew on the Fife coast. Fenton argues that Lightfoot provides an antidote to the view that the Scots spurn greens.[71]

Poaching could provide the occasional feast of salmon, hare or deer. Court records and local newspapers only very occasionally report poaching and trespass, which probably means there were few prosecutions and many a blind eye turned, except in the case of persistent – or armed – offenders. Gamekeepers, after all, lived in or near villages and had to get on with their neighbours all year round. And the poor did not regard occasional poaching as a criminal act.[72] Amy Stewart Fraser, brought up on Deeside in the 1890s, recalled that every cottager had a shotgun, for the rabbit stew and hare soup.[73]

DAILY MEALS

Rural workers' meals changed little across regions and over time, until the last few decades before the First World War. In Moray at the start of the 1800s breakfast consisted of oatmeal porridge with milk or beer; dinner was sowens with milk or beer; and supper kail brose with oatcakes or cakes made from mixed meal. Barley broth was served for Sunday dinner with meat in winter and butter in summer.[74] Potatoes with butter, milk or onions supplied a third of the daily food in winter. The minister of Bendochy in Perthshire wrote: 'the common people live on oatmeal pottage twice a day. It is the most wholesome and palatable of all their food, being purely vegetable. Such food makes men strong like horses and purges the brain of pedantry. It produces hardy Highlanders.'[75] Only in the south-east was meat a little more common.

Where a higher proportion of the wage was paid in kind – amongst the hinds of the south-east, for instance – meals were better and more varied.[76] Labourers fared worse. Annual budgets for three day labourers and a weaver

69 John Lightfoot, *Flora Scotica*, London 1777, vol. 1, 147.

70 Ibid., 268.

71 Fenton, *Food of the Scots*, 63.

72 Diack, 'Scottish farm labourer', 325.

73 Amy Stewart Fraser, *The Hills of Home*, London 1973, 58.

74 Donaldson, *GV Elgin or Moray*, 24.

75 *OSA* Bendochy, Perthshire, vol. XIX, 349.

76 Levitt and Smout, 'Farm workers' incomes', 161.

in rural parishes in Perthshire given in the *OSA* show that three of the families spent around seventy per cent of their cash income on food and by far the largest expenditure – nearly three quarters of the total – was on oatmeal and barleymeal. The remainder of the income paid for potatoes, onions and milk, with a very small allowance for protein in the form of cheese, butter, meat and bacon.[77] While their gardens are not mentioned, it would be very surprising if none of these families had patches of ground where they could grow their potherbs.

A Fife labourer with a wife and three children under ten, paid only in cash, likewise spent £12 on food out of an annual income of £18. £1 10s. went on milk and milk products, the rest on meal, potatoes and barley for the soup. They had a kailyard, for which, with their 'wretched house', they paid 13s. a year in rent. This *OSA* contributor wrote, in unusually strong terms, that 'the greatest evils of their situation arise from the lowness of their diet, and the wretchedness of their lodging … . The meagre food of the labouring man [in general] … is unequal to oppose the effects of incessant hard labour upon his constitution.' Disease and premature death were the consequences.[78]

Gibson and Smout point out that the *OSA* reports overall describe a situation where 'gently rising incomes were providing a choice in everyday diets. Generally, however, these were reported only from favoured localities where agricultural improvement, proximity to a large town or where the linen manufacture pushed up wage rates and earnings, and where present plenty was usually contrasted with past scarcity.'[79] Tea, wheaten bread[80] and butcher meat were bought by labourers in parishes near Edinburgh, Perth and Forfar, but rarely elsewhere. Unlike in the north-east later in the century, the single servants of the south-east, the bothy and chaumer men (the latter fed in the farm kitchen), also ate well at this time, with porridge with milk for breakfast; broth with pork, mutton or beef and barley bread for dinner; and potato broth or bread with butter or cheese for supper.[81]

By the 1830s and 1840s little had changed in the nature of the diet: the same vegetables; milk, meal, kale and potatoes still universal; still very little meat or fish. One significant change was in pork and bacon: the rapid spread of potato growing had stimulated a rise in pig-keeping amongst cottagers and nearly every rural family raised at least one pig a year.

> Each family feeds a pig or two, and sometimes three in the year, the flesh of which, with the garden stuffs, forms a very comfortable and wholesome meal

77 Fenton, *Food of the Scots*, 254–6.

78 *OSA* Auchterderran, Fife, vol. I, 454–5.

79 Gibson and Smout, *Prices, Food and Wages*, 233

80 Enid Gauldie notes that the limited part wheat flour played in the diet until the late nineteenth century meant that most rural Scots were not affected by adulteration (*The Country Miller*, 6).

81 Robert Kerr, *GV Berwickshire*, Edinburgh 1809, 422.

for dinner; porridge and milk are the common breakfast of males, and tea and oat-cakes or wheaten bread and butter, of females; tea is generally used in the evening; and for supper, either porridge and milk, or herrings and potatoes; cheese is often used in the evening, and at dinner and supper.[82]

The two Statistical Accounts allow a comparison of diets over forty years in the Border parish of Kelso. In the 1790s 'one and all of them live[d] much upon hasty pudding [oatmeal porridge], and boiled potatoes with milk; without deviation, they all breakfast or sup upon one or the other. Most of the adults eat of this food, at a meal, from 6 to 8 English pounds weight, including milk.' Flatulence was a common complaint.[83] In the 1830s the peasantry of Kelso were still mainly subsisting on meal, milk and potatoes, with pease and barley meal bannocks. However, they now also ate pork and herrings.[84] Tradesmen and their families, with greater disposable income, ate better. Salt meat or fish for dinner every day was common and eggs and cheese, with porridge still for breakfast and supper, was usual at least in the south-east.[85] But Alexander Somerville's miserable experience as a labourer in the garden bothy at the world-renowned Dickson's Nursery in Edinburgh in the 1830s was of a small amount of porridge with 'soor dook' ('a kind of rank buttermilk peculiar to Edinburgh') for breakfast; potatoes with salt and sometimes herring for dinner; and oatmeal with more soor dook for supper. Meat – never; wheaten bread – seldom.[86]

HUTCHISON'S SURVEY

In the second half of the nineteenth century 'the first noticeable feature in the Scotch rural dietary, which must strike even the most casual observer', wrote Robert Hutchison of Carlowrie in 1869, 'is its uniformity in most of the districts throughout the country. Whether inland or sea-coast, highland or lowland, the great staples for sustaining life amongst the peasantry of Scotland are oatmeal and milk.'[87]

Robert Hutchison of Carlowrie was a prominent landowner, merchant and arboriculturist with an estate near Edinburgh, and set out in his survey of rural workers' diets to prove beyond doubt the superiority of the Scottish diet. This had already been the conclusion of Dr Edward Smith a few years earlier in his 'hasty glance' at the Highlands and two Border counties, part of his large study of

82 *NSA* Cameron, Fife, vol. IX, 308.

83 *OSA* Kelso, Roxburghshire, vol. X, 594. This contributor was the town's physician.

84 *NSA* Kelso, Roxburghshire, vol. III, 326.

85 Fenton, *Food of the Scots*, 351.

86 A. Somerville, *Autobiography*, 101.

87 Hutchison, 'Dietaries of Scotch agricultural labourers', 4.

workers' diets throughout the UK.[88] Hutchison set out to investigate the matter more thoroughly, using Smith's analytical tools. He collected the diets of fifty-six workers from all over Scotland – shepherds, ploughmen, agricultural labourers (including farm servants), railway workers and Dumfriesshire lead miners. This was a small sample and the number of his respondents who agreed to tell him about their weekly food consumption was even smaller: many regarded 'the process of weighing and measuring their food with superstitious awe and fear'.[89] Smith had encountered the same terror in Scotland: 'the beads of perspiration would stand upon [their] brow[s] ... and nearly all gave their answers in very low tones.'[90] Natural caution on the part of the Borders labourers underlay this attitude, rather than superstition, he believed.

Hutchison's Report reveals that, thirty years after the *NSA*, farm workers were by and large eating the same food in the same quantities. In the fifty-six detailed dietaries, meat, usually pork or bacon, was eaten daily by almost every family except for the very poorest – but in small quantities. Broth was also part of nearly all the diets, with the exception of Highland families. Hutchison also found, as had Smith for England, that the man very often ate better than his family. Most often, he had a hearty breakfast with oatmeal while his wife and children took bread and tea – and the same for supper. He also often ate far more fish and meat than the rest of the family.[91]

Examples from this survey from Aberdeenshire, Fife and Roxburghshire, given below, illustrate Hutchison's contention that there was very little regional variation. Oatmeal consumption appears to have varied according to income, not district. Porridge or brose was the usual fare for both breakfast and supper, and oatcakes sometimes too. Bread with bacon, cheese or butter or by itself was now eaten at every meal, most often at tea and supper and not only – as formerly – by the wife. There were examples of undernourishment: four families in Kingsbarns in Fife headed by single mothers working as labourers had no supper, only tea with bread and butter. For ploughmen's families, although the differences in money wages were considerable, their daily food was remarkably similar – and frugal, as were the meals of the better-paid railway worker. There were plenty of vegetables in these meals, and a lot of milk, while the large helpings of carbohydrates were flavoured – it could be no more than that – with fish, bacon, cheese and butter.

88 'Report on the food of the poorer labouring classes in England' by Dr Edward Smith, in *6th Report of the Medical Officer of the Privy Council*, 1863, Appendix, 261.

89 Hutchison, 'Dietaries of Scotch agricultural labourers', 2.

90 Smith, 'Report on the food of the poorer labouring classes', 277.

91 Hutchison, 'Dietaries of Scotch agricultural labourers', 13.

Some eastern Lowland rural labourers' family meals.

8. PARISH OF ALFORD [Aberdeenshire]. Ploughman. – B.C. Family above ten years of age, 1; below, 3. Takes meals at home. Rent free. Yearly wages, £24. Wages of family, 9s. per week. Keeps pigs and poultry.

Breakfast with family, porridge and milk; sometimes tea or coffee and bread. Dinner with family, oatmeal brose and milk or vegetable broth and potatoes. Supper with family, bread and beer and cheese, or bread and milk, or potatoes and milk. Health very good.

Remarks. – The keeping of poultry is often the cause of differences between master and servants, and leads to removal from situations here. A cow is allowed in a few cases.

27. PARISH OF KINGSBARNS [Fife]. Female labourer. – J.S. Family above ten years of age, 1; below, 2. Takes meals at home. Rent £1, 10s. Weekly wages, 5s. Keeps a pig.

Breakfast with family, porridge and milk. Dinner with family, potatoes and herring, or pork broth and bread. Tea with bread and butter. Supper none. Health good.

Remarks. – Children are at school. Very industrious.

41. PARISH OF KIRKLISTON [Midlothian]. Railway labourer. – A.K. None of his family are at home. Breakfast and supper at home; takes all his meals at home on Sunday. Rent free. Weekly wages, 18s. Wife does not work. Keeps poultry.

Breakfast, tea and bread and butter. Dinner, before rinderpest made milk so scarce, he took milk and bread and cheese, now beer and bread and cheese; family, when at house, take broth and bread with beef or cheese. Tea, bread with tea, or broth, or porridge with milk, sugar, or beer; wife takes tea and bread. Health very good.

Remarks. – This labourer is employed as a 'gaffer' on the railway.

44. PARISH OF EDDLESTON [Peeblesshire]. Ploughman.- D.B. Family above ten years, 2; below, 1. Takes meals at home. Rent free. Yearly wages, £20 in money, 8 stones barley, 128 stones potatoes, fuel carted free, and 3 pints sweet milk daily. Keeps a pig.

Breakfast with family, porridge and milk., and tea and bread afterwards. Dinner with family, milk and potatoes, or broth and potatoes, and bacon occasionally.

Supper, porridge and milk, or cheese and bread and milk; family, porridge and milk. Health good.

45. PARISH OF CASTLETON [Roxburghshire]. Labourer. – R.N. Family above ten years, 1; below, 2. Takes meals at home. Rent £2, 10s. per annum. Weekly wages, 15s. Keeps cow and poultry.

Breakfast, porridge and milk, and tea with bread and butter; family the same. Dinner with family, broth and potatoes and bacon, or cheese and milk and bread and butter. Supper, potatoes and milk and bread, or tea with bread and butter. Health excellent.[92]

Table 2 Average consumption of food and drink per person per week in eight families.

Food	Consumption per adult per week
'breadstuffs' (bread, rice, flour, barley and pease)	4lb 12oz
oatmeal	6lb
fish	12oz
meat and bacon	1lb
butter	3.4oz
cheese	3.5oz
sugar or treacle	6.4oz
green vegetables	2lb
potatoes	8lb
tea	¾ oz
milk	5 ⅔ pints
soup or broth	4 ½ pints

Notes:

(a) The families came from Kincardineshire, Perthshire, Haddingtonshire (East Lothian), Peeblesshire and Roxburghshire.

(b) Two families in Crieff, Perthshire ate three times as much meat and bacon as the other families.

(c) Children under ten were calculated as the equivalent of half an adult.[93]

[92] Ibid., Appendix A.

[93] Ibid., Appendix B.

By modern standards, both Smith's and Hutchison's samples were very small, and neither author tells us how the families were selected. But they are perhaps enough, in the absence of other data, to show broad trends – in particular, significant differences between the diets of English and Scottish labourers at this time. Meat, for example, was eaten every week by virtually all the English families, but by fewer than three quarters of the Scots. Scots consumed nearly four times as much milk per week and far more tea, but less cheese. English labourers, according to Smith, ate green vegetables only in spring and early summer, when potatoes were scarce or dear. The rest of the year they had them for Sunday dinner and perhaps on one other day, although onions could be had more often, especially by the poor. In spite of the much mocked boiled cabbage tradition, he says this vegetable was only ever eaten fried, with meat, bacon or dripping.[94] In Scotland, the daily presence of broth indicates that vegetables were regularly included.

By the 1890s many workers' diets were very different. For farm servants payment in kind was dying out, but, where it survived, much of the food was sold on. In the Lothians and Fife 'gains' no longer always included oatmeal. In many ways, everyday life was becoming easier, wages were higher and farm workers' families were less isolated. The spread of the railways helped country people to get to towns and their shops more easily. And the grocers' carts, although they killed off many a tiny hamlet shop, brought wheaten flour, rice, sugar, treacle and jam to the doorstep. The cost of living dropped dramatically: between 1873 and 1893 R. Hunter Pringle calculated (from data from one shop in St Andrews) that food prices, especially prices of imported food such as sugar and tea, had fallen by about twenty per cent.[95] But what had been lost? '[Previously, there was] more porridge and milk, less tea and slops, less going to town at night, and fewer credit accounts with grocers' carts. The labourers got the bulk of their food at producers' prices.'[96]

Pringle was no less afflicted with the conservative nostalgia than most other commentators on rural life of the time, but his evidence, however slim, does show change, and deterioration. An East Lothian orraman's budget, for himself, his wife and his adult son shows they spent 15s. a week on food, buying enormous quantities of sugar (though possibly for jam-making) along with other 'new' items such as coffee essence, jam, flour, bread and rice, as well as meat (perhaps 5lb), fish, butter and cheese.[97] Vegetables do not appear in this budget but they had a garden: they spent 2s. 6d on garden seeds for the year. The only gains recorded were coals, but there is no mention anywhere of milk and oatmeal. A much less 'extravagant' budget from Fife, submitted by a cattleman's wife near Cupar with

94 Ibid., 243.

95 *RC Labour*, vol. III, pt II, Fife, Kinross and Clackmannan, app C, 82.

96 Ibid., 104.

97 *RC Labour*, vol. III, pt II, app, 141.

seven children living at home, shows she spent 8s. 1½d a week on groceries – they too probably had a garden and potato ground. The quantities, as much as the items themselves, are instructive (Table 3).

Table 3 Farm servant's family budget, Fife.[98]

Food item	Quantity	£ s d
Tea	½ lb	0 1 0
Butter	½ lb	0 0 7
Sugar	6 lb	0 1 0
Bread	6 loaves approx.	0 3 0
Peas, barley for broth	(not given)	0 0 3
Jam	½ pot	0 0 3 ½
Butcher meat	1 lb approx.	0 1 0
Cheese	2 lb approx.	0 1 0

Such diets were now much more akin to those of city workers and of the English working class than to those of the countryside in the past, and the reduced consumption of milk, particularly by children, worried commentators.

Even in the north, white bread and treacle at breakfast and supper were making inroads into the empire of oatmeal and kale.[99] In the Lothians, where payment in kind had survived for longer than elsewhere, the change was most striking. R. Hunter Pringle wrote:

> Porridge is the chief breakfast food where milk can be got, but I heard of many houses where tea with bread, cheese, butter and jam has supplanted porridge. There are two courses for dinner, either broth and meat or meat and pudding. Herrings, either fresh or salt are in great esteem … . Eggs too are purchased from farmers or 'egglers' … . Coffee and tea, costing 2/6 to 3/- per pound are drunk … if there be whiskey in the cupboard it is produced on high-days and holidays only. Some good housewives buy flour and bake scones. The barley and peasemeal scone seems to have altogether disappeared.[100]

The single farm servants, male and female, living in bothies or in chaumers in the farmhouse were usually fed quite different fare and much is known of it from horrified accounts by those who concerned themselves with their often appalling living conditions, and from the men's bothy ballads. While Hutchison in the 1860s noted that bothymen were the best fed as they were most likely to eat meat every day, for most later on, according to the men themselves, daily meals were an 'infinity of brose: brose when they fell out of their beds in the morning,

98 *RC Labour*, vol. III, pt II, Fife, Kinross and Clackmannan, app C3, 82.

99 Carter, *Farm Life in Northeast Scotland*, 136.

100 *RC Labour*, vol. III, pt II, Edinburgh and Haddington, 119.

Plate 22 Chaumer meal, no date.

kail brose for their "denner" – the mid-day meal and oatmeal brose again for their supper'.[101] Water brose, ale brose, cabbage and turnip brose might ring the changes and the men could buy their own food. A red herring or piece of meat might appear on a Sunday (Plate 22).

Monotony was not a complaint, although the Carse of Gowrie ploughman followed up a eulogy to brose ('perhaps the most wholesome and substantial food in this or any other country') with a modest suggestion that 'an occasional change of diet would not materially damage the appetite of the rural hind'.[102] Quality was another matter: the ingredients might be old, sour or stale, or the cooking done carelessly by the mistress or a woman servant – though often by the men themselves. And there was bitterness in the knowledge that the master and mistress were dining on much better fare:

> Dinna gang to the Barns if ye wish to be weel,
> A' the days o' your life ye ne'er saw sic a chiel,[such a fellow]
> He'll treat ye to a breakfast o' buttermilk and meal
> Wi' a drink o' sour ale in the mornin'.

[101] David Kerr Cameron, *The Ballad and the Plough*, London 1979, 103.
[102] *The Clodpole*, 3–4.

We get beef-bree whiles weel seasned wi' reek,[smoke]
Wi' three seeds o' barley and the smell o' a leek,
If your nae pleased wi' that, he'll tak nane o' your cheek
But he'll pit ye frae his toon in the mornin
The Barns o'Beneuchies[103]

Miners' meals seem to have been very like those of farm workers, with more meat and apparently fewer vegetables. There is one account of miners' diets, by the radical American journalist Kellogg Durland, who worked in a pit and boarded in a typical overcrowded miner's cottage in Fife in the early 1900s. For breakfast he had tea, or 'switched' egg and milk, and bread. A piece with jam or cheese, chased with cold tea, followed at 9am; dinner at the end of the working day was at 3pm. Three days of the week they had porridge and three days there were meat or eggs and potatoes. 'There was always an ample supply of bread and scones. Tea was simply tea, bread, scones and jam, and supper was just the same as tea without the jam.' Scotch broth appeared every other week and there were boiled eggs for breakfast on Sundays. 'Save for the monotony of it, this was wholesome fare.' Durland estimated that the cost of his board was at maximum 2s. 6d or 3s. per week.[104] This diet seems to have been almost entirely vegetable-free but we know that many miners grew vegetables, especially in Kelty.

A HEALTHY DIET?

As in many other Northern European countries, climate and poverty can be held largely responsible for the restricted range of foods in the traditional Scots diet. Nevertheless, it was a widely held belief that the rural working class was healthier and longer-lived in Scotland than in almost any other part of Great Britain, and that it was at least in large part their diet that made them so.

The basic diet of oatmeal and milk was notably high in iron and calcium and adequate for protein. Gibson and Smout calculate that just 20oz (the daily allowance in institutions such as prisons) would provide sixty-eight per cent of the calories, sixty-six per cent of the calcium and 100 per cent of the iron, vitamin B and niacin required by a very active adult male.[105] Maisie Steven, whose 1985 study remains the most thorough modern analysis of this diet, has calculated that the greater quantity of oatmeal consumed by farm workers, together with large helpings of potatoes and milk, would make up the rest of the calories. Potatoes, kale and turnips, 'no matter how long or how badly cooked', added vitamin C and potassium. Pease and bean meal too, mixed with barley, was rich

103 Ord, *Ord's Bothy Songs*, 281.
104 Durland, 'Among the Fife Miners'.
105 Ibid., 236.

in protein, iron and B vitamins.[106] What the diet was probably deficient in, if the milk was skimmed rather than whole, was vitamin A (causing eye disease) and enough calcium for children and nursing mothers.[107]

Dr Edward Smith concluded that Scottish farm labourers' diet was superior, mainly because of the huge quantities of milk and oatmeal they got through.[108] Hutchison's survey seems to bear this out, and showed that the Scottish diet was cheaper as well as more nutritious.[109] Even at the end of the century, when the traditional diet was being eroded, informed medical opinion held that the rural labouring population was generally better fed in Scotland than in England.[110] Scots rural workers appear to have eaten more vegetables than the English (although less meat and cheese). And this must have been because they were more likely to have gardens. Cooking vegetables in broth or soup, moreover, as was the norm in Scotland, conserves more nutrients (B vitamins and vitamin C) than can survive boiling.

Death rates in rural Scotland in the middle of the nineteenth century were significantly lower than those in Scottish towns and cities.[111] A large part of this has to be due to diet, although housing, clothing, hygiene, disease and occupation had much to contribute. The monotonous regularity of oatmeal, milk, potatoes and vegetables appears to have had an impact on height too. An individual's height is second only to their weight as an index of physical and emotional well-being and research into records of the height of army and navy recruits by Roderick Floud and his colleagues has revealed a clear trend: before the 1840s, while the height of all recruits was rising, the tallest recruits came from rural districts in Scotland, including the Highlands.[112] Floud, Wachter and Gregory argue that army recruits at this time can be taken as representative of the lower classes as a whole and so their findings can be applied to the whole of this stratum of society.[113] The most plausible explanation for the height difference is unlikely to be genetic and is most likely to be down to diet. After

[106] Steven, *Good Scots Diet*, 50.

[107] Gibson and Smout, *Prices, Food and Wages*, 236.

[108] Smith, 'Report on the food of the poorer labouring classes', 17.

[109] Hutchison based his analysis on the carbon and nitrogen content of food, as measures of its nutritional value. The recommended intake for an active 'hard-worked' British labourer, according to Lyon Playfair, was 43,793 grains per week: the average Scottish labourer's diet gave him 51,809 per week (Hutchison, 'Dietaries of Scotch agricultural labourers', 7–8); the English labourer's 42,207 (Smith, 'Report on the food of the poorer labouring classes', 291).

[110] T. Oliver, 'The diet of toil', *The Lancet*, 29 June 1895, 1629–35, 1634.

[111] P. Greaves, 'Regional differences in the mid-Victorian diet and their impact on health', *Journal of the Royal Society of Medicine Open*, vol. 9, no. 3, 8th March 2018, 1–6.

[112] Roderick Floud, Kenneth Wachter and Annabel Gregory, *Height, Health and History. Nutritional Status in the United Kingdom, 1750–1980*, Cambridge 1990, 198.

[113] Floud et al. point to the minimum height regulations and the security of employment and income that the army promised (Ibid., 200–2).

1840, while the average height of all recruits fell,[114] those from the countryside in Scotland continued – unlike in England and Wales – to be taller than their urban comrades.[115] Until the 1840s, too, the life expectancy of those born in Scotland between the 1790s and the 1840s was higher than in England.[116]

More evidence that country people in Scotland were taller comes from a study of male working-class prisoners in Glasgow prisons. Men born in the countryside before 1850 were on average 1.1 inches taller than their urban counterparts.[117] Again, access to fresh milk and vegetables, along with less overcrowded housing, outdoor work and lower disease rates undoubtedly produced this striking difference.[118]

In the 1860s the health and longevity of the rural population was still something to be marvelled at: 'it seems to be established as a fact that in no country of Europe are the inhabitants more distinguished for healthiness, longevity and freedom from disease than they are in Scotland.'[119] Rather than diet, however, this author held seasonal farm work, as undertaken by weavers, shoemakers and others engaged in sedentary occupations, responsible – and he may have been right for these occupations.[120]

Hunger was not banished, of course. The last prolonged deadly famine in Scotland was in the 1690s but in the following century there was dearth in 1739–41, which may have prompted the widespread adoption of the potato; in 1782–3 a terrible grain harvest caused food shortages and led to disease; high prices in 1799–1800 led to food riots in towns. Even in the 1800s there were years of high food prices and scarcity: notably 1810, 1813, 1816 and 1818. When the harvest failed it was the unskilled labourers paid not in kind but in cash who suffered the most, and who even in normal times might go hungry when illness struck or there was no work. What did they have to live on? From the mid-eighteenth century systems of relief organised by the church and the parish heritors ensured that no one died. Potatoes could save lives but in the dearth of 1799–1800, according to Ramsay of Ochtertyre, the last resort of the poor, after potatoes, was rice and maize imported from the USA.[121] In the 1840s large-scale relief programmes in the Highlands avoided a famine with distributions of grain, rice and maize, and hunger afflicted some of the very poor in the Lowlands too. In less extreme times of dearth, salt herring and potatoes were the usual fare. The continuing

114 The proportion of rural army recruits fell through the nineteenth century (Ibid., 94).

115 Ibid., 198.

116 Ibid., 292.

117 Paul Riggs, 'The standard of living in Scotland', in J. Komlos (ed.), *Stature, Living Standards and Economic Development*, Chicago 1994, 65.

118 Ibid., 63.

119 Tait, 'Physiological Distinctions', 13.

120 Ibid., 16.

121 A. Allardyce (ed.), *Scotland and Scotsmen in the 18th century from the MSS of John Ramsay of Ochtertyre*, Edinburgh 1888, 228–9.

insecurity of the food supplies of the poorest is suggested in Robert Hutchison's survey: although the average family got more than the minimum nutrition – they were 'plainly, but well-fed'[122] – there were many families who were poorly fed, and he recommended increased use of the much cheaper peasemeal and maizemeal, mixed with potatoes.[123] Oatmeal was clearly beyond their pockets.[124] Hutchison also ended his report with a recommendation that ploughmen should be given larger gardens than they presently had.[125]

Early spring and the six weeks before harvest might see even the better-off rural workers going hungry. Alexander Fenton, recalling his childhood in Aberdeenshire as recently as seventy years ago, wrote of the 'spring hunger' that forced people to gather wild leaves, such as fat hen, and roots. The six weeks before harvest was another period of hunger, when foraged plants in the form of bulbs and rhizomes, new shoots, greens and later seeds (dandelion and vetch) and some berries helped to make up the food deficit.[126]

THE DECLINE OF VEGETABLES

What can we conclude about the place of vegetables in the rural diet towards the end of the period? Vegetable consumption almost certainly started to decline from the last decades of the nineteenth century as working-class families became increasingly reliant on a mainly carbohydrate diet – white bread, black treacle and sugar. The arrival of cheap sugar towards the end of the nineteenth century had a huge impact on the diet. Sugar consumption in Scotland was notoriously high: 'The number of sweetie-shops Scotland was able to support began to be a source of amazement to foreign visitors', writes Maisie Steven.[127]

Bread and jam or treacle and tea could be provided quickly and cheaply, 'free[ing] the wage-earning wife from one or even two meal preparations per day, meanwhile providing large numbers of calories to all her family'.[128] At the beginning of the twentieth century an average farm servant's family in Aberdeenshire might still be drinking plenty of milk and eating porridge and brose, oatcakes, vegetable broth, potatoes, cheese, smoked haddock and Scotch broth with beef on Sundays (eggs occasionally and bacon from the pig too). But

122 Hutchison, 'Dietaries of Scotch agricultural labourers', 14.

123 Hutchison found differences in the death-rate in Borders parishes that matched the consumption of peasemeal mixed with barley meal: parishes where this was on the menu had a lower death-rate than those where it was unknown. Ibid., 2.

124 Hutchison did not go as far as Edward Smith, who encouraged the better off to donate used tea leaves and coffee grounds to the poor (Smith, 'Report on the food of the poorer labouring classes', 18).

125 Ibid., 13.

126 Fenton, *Food of the Scots*, 61–5.

127 Steven, *Good Scots Diet*, 82–3.

128 Mintz, *Sweetness and Power*, 130.

they also had white bread and syrup, treacle or home-made jam two or even three times a day.[129] The north-east nevertheless remained somewhat of a bastion against the onward march of the 'industrial' diet. Elsewhere, even where pigs and poultry were raised, much of the pork, butter, cheese and eggs, as well as some of the oatmeal, was sold rather than eaten at home. Eggs were usually sold when dear and eaten when cheap.[130]

It was the children who were mainly, and increasingly, fed on white bread and treacle, while the main wage-earner very often had a more balanced and energy-rich diet. Nevertheless, by comparison with those of urban workers, this diet was still relatively healthy and nutritious. Rickets was rife in urban slums, along with stunted growth, anaemia and scurvy: margarine, contaminated and watered-down milk, little fruit and few vegetables made for an urban diet 'seriously lacking in every one of the essential minerals and vitamins'.[131] Amongst farm workers' children there was no rickets.[132]

The fact that nearly all rural families had gardens meant that potatoes, cabbage, kale and turnip could be eaten every day and all year round – although it is worth noting that not all the families in Hutchison's survey apparently ate greens. But how many cottagers ate other vegetables – onions, leeks, carrots, parsnips, peas and beans – as seems to have been the case in England? Accounts of the everyday meals of the ordinary people rarely mention any other greens apart from kale and any other root apart from potatoes and turnips.

Easier access to shop-bought and often poor-quality food, compounded by the decline of interest in the garden and the decline of payment in kind, led inevitably to a falling-off in the nutritional value of the rural diet for farm servants and labourers. Estimates of the calorie intake of Lowland agricultural labourers in 1924 showed that it was well below what was considered, at the time, sufficient for adult males engaged in heavy work.[133]

---

129 Diack, 'Scottish farm labourer', 323.

130 Ibid., 323.

131 Steven, *Good Scots Diet*, 94.

132 Ibid., 324.

133 Smout, *Century of the Scottish People*, 128–9.

CHAPTER 7

COMPETITIONS AND SHOWS

Introducing a round-up of local newspaper reports of British flower shows, J. C. Loudon wrote in the *Gardener's Magazine* of 1836:

> The most remarkable feature of the horticultural societies of the present year is the increasing attention which has been paid throughout the country to the exhibitions of the poor. The produce of the cottagers' gardens has been generally admired; and the influence which these exhibitions are likely to have on the condition of the cottagers is most important. It is gratifying to see the rich sympathising with the poor and this must in the end, lead to the moral and intellectual elevation of the former.[1]

Was the last word a misprint? Or a deliberate jog to the elbow of the rich to remind them of their responsibilities? This chapter tells the story of the very unequal partnership between the rich and the poor that underlay the best-kept garden competitions and cottage garden flower shows in the nineteenth century.

In Scotland, no less than elsewhere in the British Isles, it was in the 1830s that the much-praised 'spirit of emulation' descended on cottage gardeners in the form of best-kept cottage garden competitions and cottage garden societies. Mostly initiated, promoted and sponsored by the rich, they survived and prospered because of the enthusiastic participation of the poor. Every county and in some counties almost every village saw keen and expert cottage gardeners running societies and organising flower shows. Even best-kept garden competitions, originally pet projects of landowners, were taken up by cottage garden societies and developed by them in imaginative ways. In the early decades the hand of the minister and local landowners is very visible. Later in the century, as tradesmen and artisans began to take leading roles in the societies, their masters appear to have taken a step back, with one notable exception: in the mining districts the guiding hand of the coal owner never seemed to let go.

[1] Anon., 'Provincial horticultural societies', *GM*, 1836, 643–75, 643.

Press coverage of the competitions and the shows, both local and national, was extensive (although very uneven[2]), running to several columns and even whole pages during their 1880s heyday. Large shows became holiday events in the rural calendar, attracting hundreds and sometimes thousands of visitors from far and wide. Smaller village shows became festivals and we know that whole streets and groups of neighbours would enter flower shows and best-kept garden competitions. Although difficult to assess in any way objectively, their impact on cottagers' gardening knowledge and skill was frequently said to have been considerable and not only on the minority of gardeners who took an active part.

BEST-KEPT COTTAGE GARDEN COMPETITIONS

Belief in the efficacy of prizes had had a grip on improvers and government since the early eighteenth century. Premiums, as they were usually called, had been offered by local agricultural societies since the 1720s for improved husbandry practices and at livestock shows. From 1784 the Highland and Agricultural Society of Scotland (HASS) awarded its own premiums for these. So, when Patrick Neill in 1813 urged landowners to offer prizes to cottagers to persuade them to improve the appearance and the productivity of their gardens, his words fell on well-fertilised ground.[3]

It was HASS which launched competitions for cottages in 1827. They were open to cottagers paying £5 or less in annual rent or occupying cottages of that value. Premiums were to be awarded 'to excite the attention of cottagers to keeping their cottages near and clean'.[4] Prizes for cottagers' bees were also announced. The following year the garden was included and sometime in the 1830s a separate prize for the garden was added. Two sovereigns (£2 sterling) were awarded for the best-kept cottage, one sovereign for the second best-kept cottage and one sovereign for the best-kept cottage garden. The judges inspected the cottages between 12th June and 12th August each year, looking for sufficiency and neatness of the fence; cleanness of the ground and neatness of the walks; quality of crops, and general productiveness of the gardener; and choice of crops. '[M]uch advantage', declared the Society, 'is derived in some districts of Scotland, from Cottagers cultivating, besides the more common crops, early potatoes, early pease, cauliflower, lettuce, with some gooseberry and currant bushes, and a fruit tree trained against the wall etc.'

2 *The Elgin Courant and Morayshire Advertiser*, for example, reported on only the 'county' Elgin and Morayshire Horticultural Society shows until 1866, when very brief mentions of village shows appeared. Fife newspapers reported on almost every show throughout the period, and *The Scotsman* was not far behind.

3 Neill, *Gardens and Orchards*, 27.

4 Anon., 'Premiums offered by the Society in 1829, VII Cottages', *THASS*, 1829, appendix 1, 38.

These criteria remained unchanged throughout the period. The paramount requirement for fencing and neatness – the first two conditions – reflects landowners' concern for the *appearance* of cottage gardens. The Society was also unable to resist the temptation to try to develop taste and an appreciation of beauty amongst the cottagers: 'in awarding the Cottage premiums, the preference will be given to those who, in addition to these requisites, have displayed the greatest taste in ornamenting the exterior of their houses, with the ground in front and at the gables.' Quality and productivity took second place. At its 1831 AGM the Society was pleased to note that cottagers were responding to their call for ornamentation: 'while an examination of the interior of the cottages ... was highly satisfactory, their exterior decoration by some flowering shrubs and plants raised in the little plot or garden adjoining was seldom neglected.'[5]

The scheme depended entirely on local landowners and their willingness to organise the competitions. To spread the burden, there was a complicated arrangement whereby only four counties and ten parishes in each county took part each year. At first cottagers responded with enthusiasm. The 1835 AGM heard from the secretary that around 600 cottagers had entered the competitions across the whole country and that 'if anything is more gratifying than another, it is the circumstance, that on no previous occasion had such an emulation been evinced by the industrious cottagers to obtain the reward offered by the Society for neatly kept cottagers and the little gardens attached to them.'[6]

The number of entrants was never large, however, and the Society was soon struggling to keep it going. In 1836 the rules were changed to encourage more entrants. Some districts were not interested and in others, apparently, the cottages were in such a bad state that they could not be entered.[7] The Society also realised that holding the competition in an area as large as a county meant that the chances of any competitor knowing who else was entering were so slight as to seriously enfeeble the essential 'spirit of emulation'. Moreover, it had become clear that limiting a county competition to a single year meant that word did not spread among the cottagers until it was too late for many to enter. From now on districts made up of four parishes would run the competition for four years. At the parish level, local landowners would, potentially, be spurred to take more interest in the competition and 'induce' their cottagers to join in. To encourage them each district – that is, the local sponsors – had now to match the Society's contribution of £14 to the premiums as proof of the 'interest which the gentlemen connected with it take in the matter'. At first, this change seemed to work, but the proprietors' apparently revived enthusiasm proved short-lived. Twelve years later, it was being proposed

5 *The Scotsman*, 9th July 1831, 3.

6 Ibid., 24th January 1835, 4.

7 Anon., 'Cottage premiums', 203.

> that the garden premium still should be one pound[8] but as this prize has failed to excite the interest among the peasantry which it so highly deserves, the committee suggest that a medal shall be also presented to the winner. The cost of a silver medal is inconsiderable, and there is no doubt that the honour of acquiring a prize, which can be exhibited to his neighbours, and transmitted to his children, is generally more valued by a competitor than double its intrinsic value in money.[9]

It was also clearly implied that the gentry needed to be reminded of their critical role in promoting the competitions: district competitions had doubled since the previous year (this apparent contradiction is unexplained) and the convenor of the Fogo, Berwickshire, committee was particularly commended for showing 'great zeal in encouraging the competition both in that parish and in Libberton, where the premiums were some years ago in operation, and ... [for] every year generously [doubling] the sums given by the Society'. As a result, 'beautifully clean cottages and tidy and well-kept gardens' had been admired at Fogo, as had 'the spirited manner in which [the inhabitants] have come forward as competitors'.[10] Fogo was a very small agricultural parish with a population of around 400, but fifteen cottagers took part in that year's event.

In 1842 an exceptionally large total of 107 cottagers from sixteen parishes had taken part,[11] but the 1854 competition attracted the lowest number of cottagers ever. In 1863 only fourteen parishes applied to run the competition and only nine competitions actually took place. What was behind this falling off? Was it the reluctance of the cottagers or the unwillingness of the proprietors? Both were blamed. In 1854 the AGM was told that 'the proprietors did not seem to feel inclined to take the trouble of setting the machinery in motion',[12] while the 1864 AGM heard that cottagers objected to the cottage inspections (cries of 'hear!'), although garden inspections were not resented. A solution was proposed: fewer objections might be raised if 'the ladies of the district' undertook the inspections.[13] It is possible that landowners resisted the inspections, which would quickly establish the true state of their workers' dwellings. Although the rules were changed around the 1870s to include garden produce and the best-kept flower plot, a slow decline had already taken hold: in 1870 twenty-two parishes in all took part, and the 'small amount of

8 The average agricultural day labourers' weekly wage in the 1840s was 9s. 12d so it may not have been the prize, more than two weeks' wages, that was the disincentive (Levitt and Smout, 'Farm workers' incomes', 175).

9 Anon., 'Cottage premiums', *THASS*, 1849, 203–10, 206–7.

10 *Kelso Chronicle*, 28th August 1846, 6.

11 *The Scotsman*, 18th January 1843, 4.

12 *Caledonian Mercury*, 12th January 1854, 4.

13 Ibid., 21st January 1864, 3.

interest' was regretted by the Society;[14] thereafter these competitions never seem to have regained their early popularity.

In its early years the HASS scheme did make an impression. In 1839 it attracted press attention when no fewer than five cottage gardens on the Moncrieffe estate near Perth were awarded prizes. The cottages were part of a row of six newly built semi-detached[15] dwellings with long gardens in front. The story of the Craigend gardens was told in the *NSA*:

> *Cottages.* – At Craigend on the Edinburgh road, about two miles from Perth, the Moncrieffe family have erected a number of elegant cottages, which have a certain space of ground for gardens attached to them. They are built with projecting eaves somewhat in the English style. Three years ago, the Highland Society agreed to award ten prizes to such cottagers in the county of Perth, paying L.5 of yearly rent or under, whose cottages should be distinguished by general neatness and cleanliness, and whose gardens should be laid out and kept with a superior style of tastefulness and elegance. The inhabitants of five of the Craigend cottages became competitors for these prizes, and the first five of the prizes held out to the whole country were adjudged to them.[16]

It is hard not to believe that the proprietor of these new model cottages and his head gardener was not somehow involved or that there was not some co-operation amongst the cottagers of Craigend. The head gardener was Robert Sadler, whose oldest son, John, became a highly regarded botanist and curator of the Royal Botanic Garden in Edinburgh. The cottages are still standing, and still have their well-tended gardens on display to the passer-by, although the M90 now bypasses them (Plate 23).

The Society's hopes were not always quite so satisfactorily met. In Currie, Midlothian, the HASS scheme engendered only 'bad passions, [while] an unwholesome jealousy and dissatisfaction rankled in the minds of the unsuccessful competitors'.[17] In spite of this demoralising atmosphere, however, Currie founded a horticultural society only a few years later.

Why the HASS scheme received such a half-hearted response is something of a mystery. It was greatly admired and inspired imitation by many landowners and some mine owners who organised their own competitions throughout the century. Best-kept garden competitions were also run by horticultural and cottage garden societies from the 1820s, and were even tacked on to ploughing competitions. They were a boon for parishes: the minister of Knockando rued

---

[14] Proceedings at General Meetings: General Meeting 22nd June 1870, *THASS*, 1871, 106.

[15] Although the Census gives ten households, there are only eight cottages there today, and no remains of any others.

[16] *NSA* Dunbarny, Perthshire, vol. X, 824.

[17] *NSA* Currie, Midlothian, vol. I, 551.

Plate 23 Craigend cottages and gardens today.

their absence: 'if premiums were to be occasionally awarded by the proprietors for the best kept cottages and neatest gardens on their estates, the advantages would be very great, there being much room for improvement in cleanliness.'[18]

Patrick Chalmers MP, Sir James Carnegie of Southesk, Captain Ramsay of Arbeadie and Strachan and Colonel Campbell of Blackhall were some of the landowners who ran best-kept cottage competitions on their estates in Angus, Kincardineshire and Aberdeenshire in the 1840s and 1850s. The earl of Home held cottage garden competitions in several villages on his estate at The Hirsel in the Borders.[19] Competitions appear to have been most common in the most rural districts, where the influence of the local landowner was strongest – the north-east and the Borders.

While the appearance and neatness of the garden were paramount, flowers were not always considered essential, at least by the entrants. The gardens entered in Patrick Chalmers of Auldbar's competition on his estate near Brechin were greatly admired in the *Montrose, Arbroath and Brechin Review* report. Nevertheless, although the crops were excellent and showed an improvement on previous years:

18 *NSA* Knockando, Moray, vol. XIII, 82–3.

19 *Berwick Advertiser*, 9th August 1880, 6.

> the chief defect with the majority [was] a want of taste in laying out, and a paucity of flowers to ornament, their grounds. It is to be regretted that the patriotic views of their generous landlord is not more successfully carried out by them in this respect: they have certainly paid much attention to the *useful*; and we are confident a little more care as to the *ornamental* would add much to his satisfaction and their own pleasure.[20]

More than twenty years later the earl of Southesk desired the judges of his competition to look only 'in the first place, to the general appearance of the gardens, then to their productions, and to the variety of the vegetables produced suitable and useful to a cottager, the soil of the gardens, their exposure, and the labour bestowed on them begin taken into due consideration'.[21] Again, no flowers and fruit or ornamentation, but a stern emphasis on utility and appropriateness (no exotic vegetables). It may be that this stricture was made so as not to exclude those – the majority? – who did not grow flowers or fruit or unusual produce and who needed their garden to grow essential food rather than as a place of beauty and repose.

Later, the Victorian passion for flower-growing became irresistible and flowers assumed their place. Prizes for climbing plants decorating the front of cottages were awarded at the Ancrum show in 1861.[22] Fetteresso Garden Competition in 1879 had prizes for 'the most ornamental house' and 'the most ornamental back of house', with several women amongst the winners.[23] In the mill village of Stanley in 1901 prizes were offered for 'flower plots decorating street doors' and the first three winners were women.[24] Flowers could be grown on cottage walls and around doorways without encroaching on the ground needed for vegetables. In 1911 the countess of Aberdeen awarded prizes, many of them again to women, for 'the best and neatest arranged box of flowers for a window (not to exceed 9" inside)' at the Haddo House Games, Horticultural, Poultry and Dairy Association Show.[25] Prizes for window boxes had originated in the soot-laden slums of English cities, so perhaps the countess was acknowledging the tiny space allowed for flowers even in country cottages. The 1911 Kelty show, in a Fife mining district, went all out for flowers, holding not only best-kept garden, flower garden and flower plot competitions but also separate contests for best beds of begonias, calceolarias, stocks, roses, violas and roses, along with beds of peas, parsley, leeks and beans and various vegetables from seed.[26]

---

20 *Montrose, Arbroath and Brechin Review*, 24th September 1849, 3.

21 *Dundee Courier*, 25th July 1877, 4.

22 *Kelso Chronicle*, 20th September 1861, 4.

23 *Aberdeen Journal*, 7th August 1879, 7.

24 *Dundee Courier*, 19th August 1901, 3.

25 *Aberdeen Journal*, 9th August 1911, 3.

26 *Fife Free Press*, 5th August 1911, 3.

As with the HASS competitions, these local events were designed for cottages valued at £5. Some were specifically designed for ploughmen or agricultural labourers. We know something about who the winners were as, where they have been named in newspaper reports (rarely in the case of HASS competitions but more often for the local competitions), they can sometimes be traced in census records.[27] From this admittedly sparse evidence the impression is gained that, on the whole, entrants were poorer, employed in less skilled occupations and older – sometimes very old. It also appears that there was very little difference between the kind of people who won prizes in the north, the Central Belt and the south-east. The poorest were mostly over forty and worked as labourers, handloom weavers, journeymen tradesmen, estate workers, shoemakers (the poorest tradesmen) and very occasionally farm servants; a surprising number were female paupers, including two of the possible Craigend winners. Better-off winners were crofters and small farmers, farm stewards, gamekeepers, shepherds and tradesmen such as wrights, master weavers and masons.

Among HASS winners in Midlothian and Berwickshire in the 1840s were a labourer, a farm servant, a carter, a tailor, a mason and a schoolmaster.[28] Women competitors were not unusual. Amongst five women identified in north-east competitions, it is not difficult to imagine that Ann Henderson, winner in the 1854 Strachan competition at Banchory on Deeside, aged forty-five, unmarried, and her brother's housekeeper on his small farm on wasteland, had the energy and resources to garden successfully.[29] But Widow Fraser, who won at a nearby competition the year before, was a pauper in her sixties living with two pauper children and a daughter who supported the whole family. How did she manage? Widow Petrie, a prizewinner in the same competition, was even older and had the care of a small grandson too. These poor old women may have been helped by their families and neighbours: this competition, organised for Captain Ramsay's cottagers on his Arbeadie and Strachan estate, attracted forty to fifty entrants and gives the impression of a big community event.[30]

The competition sponsored by the earl of Aberdeen, a conscientious landowner, at Tarland in 1880 awarded prizes to an almost complete spectrum of village society of this time: the schoolteacher, the policeman, the gamekeeper,

---

[27] The data these searches yielded cannot afford more than an impressionistic picture of the kind of people who were winning prizes. Census Enumeration Books for Scotland, giving information on individuals' age, occupation, marital status and, from 1861, the number of rooms with windows in the house, are available at the Scotland's People website, https://www.scotlandspeople.gov.uk/.

[28] *The Scotsman*, 1st February 1843, 4; Anon., 'Premiums awarded by the Society in 1848: cottages', *THASS*, 559.

[29] *Aberdeen Journal*, 20th September 1854, 3.

[30] Ibid., 5th October 1853, 3.

a van driver, a shoemaker and a labourer.[31] Agricultural labourers seem to have formed the largest group of prizewinners in most competitions but attempts to rope in farm servants[32] met with limited success. At Errol, Perthshire, in 1870 the hiring fair for farm servants and a 'miserable failure' of a horse race (only one horse ran) were held alongside the more successful flower show. Prizes were awarded for the best-kept cottage and ploughmen's gardens on clay and on 'blackland' or more fertile soil: the winners included a bootmaker and a farm overseer but no ordinary farm servants.[33] The earl of Southesk's prizes for cottagers on his estates in Angus in 1877 went to two grieves, one farm servant and nine agricultural labourers.[34] Most of the latter were living in tiny and overcrowded farm cottages: five of the households occupied only two rooms each. John Mathers inhabited a two-room cottage with his wife and three children; into William Murray's three-room cottage were packed his wife, six children and an aunt.

The Lunan and Vinney Water Ploughing Association held a best-kept garden competition alongside its ploughing competition in Angus in 1856. Ploughing competitions were hugely popular in the second half of the nineteenth century and this one attracted sixty ploughs. Two of the winners in the ploughing competitions were also best-kept cottages and gardens winners, but it is not known if they were grieves or ordinary ploughmen.[35] The farm servants may be missing because 'frequent flitting' makes them hard to trace in the census records and it is likely that it was this habit that took them out of the competitions too. These were often judged in the summer when the men might only just have moved in to their new homes. Short sojourns in tied houses, a utilitarian attitude to their plots and the remoteness of most farm cottages from villages are very likely to have discouraged their participation.

Paternalism, reputation-enhancement and neighbourly rivalry may have entered into landowners' decisions to run garden competitions for the workers on their estates as much as the desire to get them to keep their gardens up to scratch. In the Borders eighteen cottagers at The Hirsel were awarded prizes by the earl of Home in 1880. As well as eight agricultural labourers, there were a number of estate workers: the groom at The Hirsel, a carpenter, a gardener's labourer, a forester and a gamekeeper. The large number of winners and their high average age of fifty-seven suggest rewards for long service as much as for good gardening practice.[36]

---

31 Ibid., 3rd August 1887, 11.

32 Newspaper reports may not always have distinguished between these two groups.

33 *Dundee Courier*, 29th July 1870, 4.

34 Ibid., 25th July 1877, 4.

35 *Montrose, Arbroath and Brechin Review*, 4th January 1856, 3.

36 *Berwick Advertiser*, 9th August 1880, 4.

The popularity of the competitions is undeniable and their survival gives an indication of their appeal. Today council-run competitions still take place every year. On the other hand, we do not know how much pressure may have been exerted on cottagers to enter a best-kept garden contest, especially in colliery rows such as Fordell or in estate villages, as at The Hirsel.

COTTAGE GARDEN SHOWS

Cottage garden flower shows became almost instantly popular with gardeners, the public and the philanthropic upper classes alike.[37] In 1837 the Kilmadock and Kincardine in Menteith Farm and Cottage Garden Society held its first show. The *Stirling Advertiser* commented:

> It was delightful to see so many persons crowding round the tables, eagerly examining the different articles; and it was no less pleasant to see the happy faces of the different successful competitors ... . A spirit of emulation has now gone forth among the farmers and cottagers ... which, we are persuaded, will have the effect of greatly improving the cultivation of their gardens.[38]

This society was founded by the cotton mill workers of Deanston, near Stirling, with the support, and perhaps the prompting, of the enterprising manager James Smith. The show 'exceeded the expectations of the most sanguine'. Loudon's *Gardener's Magazine* reprinted the *Stirling Advertiser* report and continued to reprint reports of this society's shows (perhaps simply for the pleasure of repeating its sonorous name). This kind of publicity, especially that given by the *Gardener's Magazine*, was clearly designed to prompt the gentry and the professional gardeners amongst its readers to encourage such societies, and they did: local elites were critical to the founding and often to the success of the cottage garden societies.[39] One of the earliest cottage garden societies was the brainchild of the indefatigable factor and gardening writer Archibald Gorrie, in his own village of Rait in the Carse of Gowrie in the mid-1820s. He was supported by his employer, the landowner Mr Stuart of Annat, and by the nursery of Dicksons of Kinnoull. Gorrie recorded that he was, at the founding meeting, 'glad to observe that the regulations had only to be read to ensure a ready acquiescence'.[40] At their 1830 show the exhibits included

[37] In the absence of society records, apart from two or three, *GM* and newspaper reports have been the main sources.

[38] Quoted in 'Provincial horticultural shows', *GM*, 1837, 581.

[39] Use of the names 'horticultural society' or 'cottage garden society' was not consistent but where shows took place in schoolrooms on Saturdays, and show classes did not include glasshouse plants, it can be assumed that, regardless of the name, they were in fact for cottagers.

[40] A. Gorrie, 'Rait village library', *GM*, 1827, 374–5, 375.

'cauliflower, cabbage, lettuce, turnips, pinks, violets, nosegay, carrots, onions, gooseberries, cow cabbage or Cesarean cale, seeds from Mr Loudon; carnations, etc'.[41] Such a range of exhibits together with the celebrity support of Loudon for a cottage garden show at this time marks this as an exceptional event. The most successful winner, the society's president, David Moyes, nearly eighty years old, surrendered his first prize, for a flower-plot, to encourage the younger villagers. Another landowner-inspired show was held in the same year in Baledgarno, a tiny village on Lord Kinnaird's estate in the Carse of Gowrie, not far from Rait.[42]

Such societies had nothing in common with the much older urban florists' societies, established and run by tradesmen and small merchants and dedicated mainly to the breeding of ever fancier varieties of tulips, auriculas and carnations, which had spread across England from the late seventeenth century.[43] Records of their shows reveal that the social side – gambling, feasting and drinking – played a large part in their success.[44] In Scotland the florists of Paisley – intelligent, well-read and self-confident silk weavers – represented this kind of society. The format and the spirit of cottage garden societies owed more, however, to the county and city horticultural societies being set up everywhere, by noble landowners, agriculturists and horticulturists, following the example of the London Horticultural Society (founded in 1804) and the Caledonian Horticultural Society of Edinburgh (1809).

In an article of 1832 in *The Quarterly Journal of Agriculture*, reprinted in the *Gardener's Magazine*, James Menteath sketched out the rationale for cottage garden societies from the perspective of the landowner. His idea was for parish gardening societies exclusively for labourers, with regular shows and prizes of seed packets for the best vegetables and fruit 'and perhaps of the more ordinary kinds of flowers'. Landowners would provide cuttings, seeds and evergreen shrubs and each society would also buy seed in bulk to sell to members. The benefits that would flow would include – as well as the fostering of the 'general spirit' of gardening in the locality through the prizes and the newspaper reports of shows – an improvement in gardening techniques, the introduction of new varieties and training for young men. Local newspaper reports would spread the word and inspire new societies. The proprietor would be rewarded with the gratitude of his cottagers.[45] Although landowners might play the major role in starting such societies, wrote Menteath, the aim was to achieve independence. It is the extent of such independence that is the most difficult aspect of the cottage garden societies to determine.

---

41 *Perthshire Courier*, 12th August 1830, 3.

42 *Perthshire Courier*, 20th September 1830, 4.

43 Willes, *Gardens of the British Working Class*, 96–7.

44 Thomas, *Man and the Natural World*, 229.

45 Menteath, 'On cottagers' garden societies', 791.

Many more societies were set up in the 1830s and 1840s. The Rait and Baledgarno shows were held in late summer, when there was most produce to display, but some societies held two or three shows between late spring and early autumn. The August and September shows drew most attention and the biggest crowds. Fife quickly took the crown for the most popular shows and probably spawned the greatest number in the east of Scotland. The *Fife Herald*, which maintained its keen interest in flower shows and gardening throughout the century, remarked in 1850: 'there are few towns or villages in Fifeshire that have not their Cottage Gardening Society.'[46] 'What has our Cottage Garden Society done?' inquired a correspondent to the *Fife Herald* rhetorically: formerly neglected gardens now yielded twice the weight of produce, thanks to more care and improved techniques.[47]

The most important show in Fife was soon that held by the Markinch Cottage Garden Society in September. Markinch was an industrial village of paper mills, cotton mills, mines and a domestic linen industry, surrounded by farmland. Started in 1839, the society held its first show in 1841, and the occupations of the prizewinners on this occasion mirrored the economy of the village: among them were a paper mill worker, four linen handloom weavers, an agricultural labourer, a hedger, a molecatcher and a wright, with an average age of forty-one, reflecting the lower age of show entrants compared to best-kept garden entrants.[48] By contrast, Monimail and Collessie Gardening Society, north of Markinch, served two agricultural and linen handloom weaving parishes, but the prizewinners in 1851, nearly all inhabitants of the planned village of Letham, were mostly self-employed or small employers: two innkeepers, a blacksmith employing one man, a master house carpenter, a baker, brewer and employer of nine men, and a master gardener. A little lower down the social scale were the gamekeeper, a clerk and a shoemaker. There was only one weaver. Their average age was forty-two.[49]

South of Edinburgh, the Horticultural Society in Currie opened its first show at the parish school in August of 1849. In lovely weather and to the strains of a band 'a large assemblage of nobility and gentry' gathered to admire the prize blooms, fruit and vegetables on display in the school playground. There were separate classes for cottagers and professional gardeners. Amongst the twenty-five cottagers who won prizes were a single agricultural labourer, along with two tailors (one with a shop), a grocer and spirit dealer, an innkeeper, three farmers and a widow described as 'a life renter of houses'. At least half a dozen of these better-off winners were near neighbours in the village of Currie. Did this

46 *Fife Herald*, 10th January 1850, 4.

47 Ibid., 21st May 1849, 4.

48 *Fifeshire Journal*, 29th July 1841, 3.

49 Ibid., 17th July 1851, 4.

encourage co-operation – or rivalry? In spite of the presence of three large paper mills in the parish, no industrial workers won prizes.[50]

Some shows did make special efforts to attract farm servants and agricultural labourers. The horticultural society in the Borders market town of Kelso, whose patron was the duke of Roxburgh and president a local landowner, James Stormonth Darling, and which catered mainly for professional gardeners and amateurs (people who employed a gardener), preserves in its records a schedule for its 1850 July show. A local seedsman had donated prizes for farm servants and labourers. As might be expected, the prizes were exclusively for kale and cabbages:

> The sum of 20s, given by Mr Somner, Seedsman, to be competed for by Stewards, Hind, Shepherds, and Cottagers, under a Farmer, for Plants purchased from his Shop in Spring 1850, viz:

	s. d.		s. d.
For the two best early cabbages	3 6	For the two best Savoys	3 6
For the second best do.	1 6	For the second best do.	1 6
For the best Late or Drumhead Cabbage	3 6	For the two best Curled Greens	3 6
For the second best do	1 6	For the second best do.	1 6[51]

By the second half of the nineteenth century cottage garden flower shows had become well established across Lowland Scotland. For working-class gardeners, many with more leisure time and higher real wages than ever before, shows were a chance to show off their prowess, enjoy an entertaining day out with their families and marvel at displays of new and exotic plants from great houses and from nurseries. The prime aim of all shows, however, was to encourage improved gardening techniques and pride in gardening efforts: 'one main object of the society [Leuchars] is to stimulate the cottager to a more profitable cultivation of his garden', noted the *Fife Herald* in 1861.[52] Many shows seem to have succeeded. The *Perthshire Courier* prefaced its report of the Logiealmond and Glenalmond show in 1864 with this striking testimony:

---

50 *The Scotsman*, 8th August 1849, 3.

51 Kelso Horticultural Society Records, extract from Kelso Horticultural Show Schedule 1850, reproduced by kind permission of Kelso Horticultural Show Committee.

52 *Fife Herald*, 12th September 1861, 3.

> it is to the cottagers throughout the country, and in the village, that we would give all merited praise. Five years ago, the most sanguine could not have expected such improvements in the district, so radical has been the change. It is no unusual thing now, to see the most beautiful and rare flowers, where only docks and nettles grew before.[53]

Such extravagant admiration was not at all unusual. Cottagers could even emulate the professionals: at the September 1861 show put on by the Horticultural Society of Longforgan, Liff and Benvie, Perthshire, 'the superiority of many of [the cottagers'] productions was such as we would only expect from gardeners.'[54] The 1889 St Andrews show seems to have been boycotted by professional gardeners, or their employers, but the cottagers' exhibits were 'far above the average, evincing that in this class floriculture is becoming more and more popular every year'. At the Lasswade and Cockpen show in 1877 cottagers' phloxes and spikes (delphiniums) were better than the gardeners'.[55]

Shows could also have other beneficial effects. David Gorrie tells a story of two apprentices who stole exhibits from village gardens the day before the show, and who, instead of being punished by the committee, were helped and encouraged to become gardeners. Later both emerged as leading members of the local gardening society.[56] On the other hand, Bob Holman recalled pre-show thieving in Fordell,[57] while in Braco in Perthshire the village show had to be given up because of the 'borrowing of produce from the big hoose gardens'.[58]

THE SHOW COMPETITORS

The stories of those who took part in the cottage garden shows often reflect the social and demographic variation in the Scottish countryside and the changes that took place, particularly after 1850. Before the mid-century, rural labourers, factory operatives and handloom weavers were much more likely to appear in lists of prizewinners, as at Markinch in 1841, where, of nine identifiable prizewinners, there were four handloom weavers and four labourers.[59] Again, in 1851, most of the identified winners were low-skilled or unskilled.[60]

Other village shows were more socially mixed and at some tradesmen and small employers predominated, as at Currie and Letham. Even the mining village of Roslin's 1849 show awarded prizes to two gunpowder makers, two

53 *Perthshire Courier*, 13th September 1864, 4.

54 *Dundee, Perth, Forfar and Fife People's Journal*, 14th September 1861.

55 *Dalkeith Advertiser*, 6th September 1877, 4.

56 David Gorrie, *Passages in the Life of a Ploughboy*, Dundee 1857, 35.

57 Holman, *Behind the Diamond Panes*, 45.

58 Daniel J. McIldowie, *The Parish of Ardoch and the Village of Braco*, Braco 1981, 25–6.

59 *Fifeshire Journal*, 29th July 1841, 3.

60 Ibid., 24th July 1851, 3.

agricultural labourers, a cooper and a blacksmith.[61] The Gauldry show of 1871, serving an agricultural district in north Fife, gave prizes to a blacksmith, a weaver, joiners, a master saddler, a salmon fisher and no fewer than four shoemakers – but here a farm servant and two agricultural labourers were also among the winners. Two ploughmen also won prizes for the most tastefully arranged and best-kept garden.[62]

The ploughman David Gorrie in the 1850s understood why tradesmen were the 'most successful competitors' in flower shows. For them, unlike farm servants or agricultural labourers, gardening was a recreation: 'There is little relaxation of mind experienced in exchanging the field for the garden, but much in entering the garden from the shop, the stall or the bench. The neatest gardens in a country village are generally those of the blacksmith, the builder, the tailor, and the shoemaker.'[63] As villages grew and evolved after 1850, some becoming small industrial or commercial centres, the shows were increasingly dominated by tradesmen and business men, without, however, losing their poorer constituencies. Markinch had grown into a small and thriving industrial town by the 1880s and the flower show winners in 1880 were now mostly younger businessmen, often found entering the most fashionable pot plant classes.[64] It is easy to picture these up-and-coming men – the gas manager and the bleach-works manager, the draper and the coal merchant, all in their thirties and forties – deciding to improve their social standing by taking part in the event that marked one of the high points of the social year. Nevertheless, the society continued to call itself a Cottage Garden Society and ordinary workers – a paper mill worker, forester, postman and railway porter – were also among the 1880 winners. Postmen, railway workers and police constables – some of them probably former farm servants – start to appear amongst lists of winners in the north-east too.

In the town and city shows middle-class women took part from early on, but women rarely appear in the lists of winners in cottage garden shows much before the end of the century, and then only in flower, and sometimes fruit, window box and decorative front door, classes. A notable exception are the two women in their twenties who won prizes for vegetables at the 1871 Gauldry show – Marjory Smith, a milliner and dressmaker and Elizabeth Taylor, a housekeeper.[65] At Meigle, as late as 1891, the only females who won prizes in the very extensive flower classes were schoolgirls who had entered bouquets of wild flowers.[66]

61 *Caledonian Mercury*, 28th August 1845, 3.
62 *Fife Herald*, 10th August 1871, 3.
63 D. Gorrie, *Life of a Ploughboy*, 37.
64 *Fife Herald*, 2nd September 1880, 8.
65 Ibid., 10th August 1871, 8.
66 *Dundee Courier*, 10th August 1891, 3.

THE EXHIBITS

The preferences of the committee, local conditions, local tradition and the popularity of certain classes, and fashion too, all in various ways determined the choice of classes for a show. Cottage garden societies very often limited their classes to an unchanging list of old-established, hardy and easy-to-grow produce, and often only vegetables. This was often even more true of the cottager classes in the grander shows such as Kelso. Was this the influence of the minister and the gentry, anxious to encourage useful crops?

Vegetable classes at the earlier shows were fairly limited. In these, as in the fruit classes (and perhaps in some of the flower classes too), it was weight and size that counted. The prize cabbage at the 1849 Kilmadock and Kincardine show weighed 21¾lb.[67] Kennoway's summer show in 1837 had early potatoes, early cabbages and lettuces.[68] Markinch's 1841 show was in July and featured carrots, parsnip and three types of brassica.[69] Currie's late summer 1849 show added more, but still very humdrum, vegetable classes: beans, cauliflowers, early potatoes, German greens and savoys, spring sown onions, late potatoes and potato onions.[70] In the 1880s the vegetable classes at the big Fife shows of Markinch and Kennoway were almost unchanged from the early years apart from the addition of rhubarb and shallots. From the 1890s a wider range of crops begins to creep in and varieties start to be listed. At Meigle three varieties of potatoes (ashleaf kidney, round red and red early kidney), cabbage, early cabbage and pickling cabbage, red onions, potato onions and autumn onions and red and white celery all had their own classes.[71]Twenty years later an Aberdeenshire show had classes for marrows and parsley and seven varieties of potatoes.[72]

Until the 1850s it was unusual to find fruit entered in cottage garden shows. Encouragement was not lacking: the nursery firm of Dickson and Turnbull donated a prize to the Montrose show of 1836 for the greatest variety of apples, pears and so on entered by 'operatives or cottagers'. It was won by James Robb, a thirty-year-old linen handloom weaver from the tiny suburb of Tayock Bridge.[73] In 1849 Currie – this show does seem to have been more advanced than shows further north, perhaps thanks to its proximity to Edinburgh – offered strawberry and raspberry classes. Fife and Perthshire shows later in the century mainly featured currants, gooseberries and apples. Gauldry (1871) added cherries,

---

67 *NBA*, 26th September 1849, 317.

68 *Fife Herald*, 29th June 1837, 3.

69 *Fifeshire Journal*, 29th July 1841, 3.

70 *The Scotsman*, 8th August 1849, 11. Potato onions are related to shallots and reproduce by bulb division, so were easier to grow than seed-sown, especially in Northern European conditions. They overwinter well in the ground.

71 *Dundee Courier*, 10th August 1891, 3.

72 The show was at Cornhill. *Aberdeen Journal*, 11th August 1911, 4.

73 Anon., 'Provincial horticultural societies', *GM*, 1836, vol. XII, 671.

raspberries and strawberries; Markinch (1880) pears and plums; Meigle (1891) had red, yellow and green gooseberries and Jargonelle pears.

Geography and climate played a smaller part in the choice of classes than fashion and the committee's idiosyncrasies and perhaps the social make-up of the membership. In the Borders, Abbey St Bathans in 1880[74] and Ayton in 1911[75] still had no fruit classes at all, while the Earlston Cottages 1901 show had white, red and black currants, dessert and baking apples, plums and Victoria plums and gooseberries.[76] Cornhill, Aberdeenshire, in 1911 had the three kinds of currant as well as four kinds of gooseberry, baking and dessert apples and strawberries and cherries.[77]

Flower-growing by cottagers, especially cultivation of new and exotic blooms, was often not encouraged earlier in the century, as we have seen. Nevertheless, the earliest shows did include flowers, and not only traditional cottage garden species. Kennoway's first show in 1837 awarded prizes for mimulus, heartsease, stock, gillyflowers, roses, pinks and ranunculus – all traditional – but also the new bedding plants: calceolarias, geraniums and antirrhinums. Kilmadock and Kincardine in Menteith had 'a great variety' of the very fashionable dahlias, which also appeared in the cottagers' classes at Cupar and at Markinch. Although, but perhaps because, they demanded so much care, late-flowering dahlias (the ball-shaped variety) soon became the unchallenged stars of early nineteenth-century flower shows, even determining the date of the show.

Even the smaller village shows mixed cottage garden favourites – sweet william, heartsease, daisies, roses, phlox, marigolds, stock, carnations and pansies – with classes for the newer fuchsia, hydrangea, gladioli, penstemon and African marigolds. The miners of the Mossgreen and District Society had almost as many different vegetable classes to enter as classes for flowers at their 1880 show but not a single traditional flower: instead, French and African marigolds, globe asters, show pansies, gladioli and lobelia, and cockscomb in a pot.[78] By 1880 Markinch had forty-five classes for flowers, including picotees (a type of carnation), quilled asters, chrysanthemum asters and French and African marigolds, and only eleven vegetables classes, reflecting its increasing sophistication.[79] Shows in farming districts such as Ancrum in Roxburghshire and Meigle in Perthshire were still dominated by vegetables. Was this because the competitors were more likely to be 'ordinary' cottagers or, again, the influence of the landed interest? Meigle in 1891 had more classes for vegetables

---

74 *Berwick Advertiser*, 20th August 1880, 2.

75 Ibid., 8th August, 1911, 3.

76 *Berwickshire News*, 3rd September 1901, 7.

77 *Aberdeen Journal*, 11th August 1911, 4.

78 NRS, GD172/2847/3, Schedule, Mossgreen and District Floral and Horticultural Society Annual Exhibition, 11th September 1880.

79 *Fife Herald*, 2nd September 1880, 8.

than for cut flowers, even though the local solicitor, hotel-keeper and linseed cake merchant were amongst the winners.[80]

Raising and preparing these cut flowers for exhibition called for a great deal of care, patience, skill and time, and often special equipment. Dahlias, for example, had to be sheltered from the weather with little canvas caps, cut at exactly the right moment before the show and then carefully packed.[81]

Potted plants (geraniums and fuchsia) first seem to have appeared in shows in the 1840s. Their popularity grew rapidly. Markinch in 1880 had six kinds of geranium, six of fuchsia, two of petunia, begonia, alpines and various foliage plants – twenty-six classes in all. Dysart cottagers' classes included thirty-seven potted plant classes ranging from a single geranium to the indiarubber plant, palm, dracaena and, of course, the aspidistra.[82] Cornhill, a small agricultural village with a large show (and berry market), had classes for double geraniums, foliage plants (aspidistra again?) and a 'table decoration plant' in 1884.[83]

Did workers enter flower classes? Until the 1880s poorer competitors were more likely to enter their vegetables than their flowers. Nevertheless, at the 1841 Markinch show the hedger, Alexander Philp, won prizes for his China roses, John Main the molecatcher for geraniums, James Honeyman, one of the linen handloom weavers, for his sweet williams and George Millar, a servant at Auchmuty paper mill, for mimulus. In 1880, at the zenith of the flower show, Joseph Low, a paper mill worker, took prizes at Markinch for a tender plant, a begonia and an ornamental foliage plant.[84] Miners exhibited flowers at their shows in Fife and the Lothians. At the Kelty show in 1911 miners showed beds of violas, calceolarias and roses.[85]

RUNNING THE SHOW

Even the smaller village shows needed considerable organisation and society committees might meet throughout the year to settle on a venue, choose classes, select judges, find sponsors for prizes and ensure that professional gardeners and nurseries came to set up displays. Venues changed and grew, although the smaller village shows continued to make do with a few tables in the parochial schoolroom or playground on a late summer Saturday afternoon. The larger shows soon graduated to town halls, mechanics institutes and the grounds of great mansions, with marquees, marking the growing importance to communities of the flower show.

---

80 *Dundee Courier*, 10th August 1891, 3.

81 C. H. Curtis, *The Book of the Flower Show*, London 1910, 45.

82 *Fife Free Press*, 17th August 1901, 5.

83 *Banffshire Journal*, 19th August 1884, 4.

84 *Fife Herald*, 2nd September 1880, 3.

85 *Fife Free Press*, 5th August 1911, 6.

Over the years more competition classes were added, including honey, poultry, pigeons, caged birds, rabbits, cats and dogs. In the 1880s 'industrial sections' for women start to appear, with classes for jams and jellies, baking, knitting and sewing. Many shows became more elaborate and more fun. Music and dancing, football matches and a complete sports programme became common. At the 1911 Wemyss Floral Fete, in a mining village, there was a band contest with eight colliery and other bands entering from across central Scotland.[86] Such attractions made for a fairground atmosphere, which helped to draw the crowds, and these were often enormous – thousands were not uncommon.

Fife had more shows than any other county. Here, in the 'Kingdom', the many smaller shows in farming areas, industrial villages, colliery rows and fishing villages were outshone in pomp by the great village and small town shows, most famously in Kennoway and Markinch. They remained focused on cottage gardens, even though classes for amateurs were often added and many societies had assumed the name 'horticultural' rather than 'cottage garden'. Held on a Wednesday in late summer, these were all-day festivals with exhibits displayed in large marquees (the huge Fife Marquee spent the season being hauled from show to show), bands, sweetie stalls lining the streets and dancing in the evening. A pipe or instrumental band became essential, and military bands were in great demand, Markinch hiring one from Edinburgh Castle. Societies competed to lay on the largest show and the *Fife Herald* faithfully covered them all – more extensively than any other newspaper. In the late summer of 1888 a total of seventeen shows was reported to have been held in Fife.

The 1880s saw the high water mark of the flower show. *The Scotsman* listed thirty-five shows in the Lothians and the Borders in the summer of 1888. As might be expected, Fife excelled itself (even in the titles of its shows). Kennoway, a typical Fife village of agricultural workers, handloom weavers and shoemakers, much smaller and less prosperous than Markinch, had been running annual shows since 1836. In 1877 more than 2,000 exhibits had been entered for the cottagers' classes alone.[87] The *Fife Herald* report below gives a flavour of the local pride, competitiveness, excitement and respectable pleasure-seeking the best village shows thrived on:

Kennoway's 'Grand Annual Flower Show and Musical Promenade'

The Kennoway Floral and Horticultural Society was the first in Fife that had the enterprise to hold their annual show in a marquee erected in a picturesque situation, and to summon a band of music to enhance the pleasures of the occasion by discoursing sweet sounds. Many other Societies in 'the kingdom', with more advantages in their favour have followed the example of Kennoway, and have no doubt drawn larger crowds of sight-seers; but none, not even Markinch, has

[86] *Fife Free Press*, 12th August 1911, 3.

[87] *Fife Herald*, 16th August 1877, 3.

ever been able to eclipse Kennoway in the superiority of the articles produced for competition, whether flowers, fruits, or vegetables. This year the show was no exception to the rule. In glancing round it, and reflecting on the unpropitious character of the season, one was astonished to find so many beautiful specimens of the products of the garden. But in the cottagers' competition and in the open competition, remarkably fine fruit was exhibited, while in the section for vegetables, rhubarb of giant size and magnificent potatoes were there in abundance; but the cottagers and amateurs of Kennoway are no mere utilitarians, cultivating only what will delight the palate or fill the stomach. They excel equally in flowers – the poetry of nature. Some of the pot plants, particularly geraniums, fuschias [sic], hydrangeas, and ferns would have done credit to professional gardeners with all their appliances, glass-houses, and facilities for heating and watering.

The show was held in the spacious Fife Marquee which was erected on the level portion of the Den. Tables were ranged along the sides, while a broad table ran up the centre. All these tables were well filled. Outstanding features on the centre table were the floral devices and model flower gardens. The first prize for floral device was gained by Mr. P. Mann, who exhibited a highly elaborated tower of Chinese pattern, with real water fountain playing at the foot, among miniature rocks and flowers. This was a genuine work of art, betokening not only painstaking and skill, but originality in design and taste in the harmonising of colour. The second prize was gained by Mr Wm Brown, who had a device of similar character, though not quite so elaborate, but very creditable to the artist. Mr D. Philip, Falkland, gained the third prize with a device representing the famed Palace of his own Royal burgh. Exquisite model flower gardens were exhibited by Messrs James Shand, C. Normand, and Robert Taylor, who gained the three prizes respectively in the order named. They were all good, but Mr Shand's was quite a gem. In addition to these the centre table was enriched by palms, ferns, fuschias, hydrangeas, and coleuses from the gardens of Kilmux, Cameron Bridge House, and Kingsdale. Some beautiful stands of violas and roses were also sent from Balcarres.

Leven Brass Band, in their cheerful uniform, played a large number of popular airs, occasionally adding choral richness to their excellent instrumentation. A stand was, as usual, erected for them on the other side of the stream under a sylvan bower, which seemed to act as a natural sounding board, and, with the running brook at their feet, imparted a mellifluous effect to the musical cadence. When the crowd assembled in their holiday attire, and moved about or squatted on the grass, the scene presented in the Den was such as might well inspire the genius of the 'sweet singer of Kennoway' or any other favourite poet. Toward evening the lads and lasses 'tripped it' on the green to the music of the band, and the utmost harmony and innocent joy prevailed till after dark when, it being now nine o'clock, the company joined in singing 'Auld Lang Syne' which was the signal for finishing the day's amusement. Fortunately, the weather, though sometimes threatening, remained good throughout. In the evening when the lamps on the 'sweetie stands' were lighted, the young people the while enjoying their dance, and crowds looking on, the effect was very picturesque and striking.

> The show had drawn together a large number of sweetie vendors, showmen and others, strangers to this village; but neither among them nor among the local frequenters of the exhibition was there the slightest disturbance. All passed off peacefully and successfully. The amount drawn for admission was, we are glad to learn, somewhat over the sum realised.[88]

Amongst the Fife shows, however, the best-known was Markinch. Every September Markinch show drew the greatest crowds and attracted the greatest media hyperbole. For weeks before the event the local newspapers carried advertisements for it, together with titillating news items about the special trains organised and the five-foot leek being nursed up in Cupar.[89] The 1880 show took place on Wednesday 1st September in the grounds of Balbirnie House, the seat of the wealthy Balfour family, in 'delightful weather'. Works and factories closed early[90] and 12,000 flocked to this show, carried on the special trains 'while carts, cars, coaches etc dashed along the roads, all bound, with their living loads, for the attractions of Balbirnie, the beautiful grounds of which were open to the visitors, who enjoyed a rich and rare treat' (Plate 24). 'Verily', concluded the ecstatic reporter, 'Markinch may well crow'.[91] As usual the exhibits were much improved since the previous year, as well as being superior to those of all other shows.

Displays of exotic flowers and produce from the gardens and greenhouses of local mansions and from commercial nurseries were an important element of the show, even at the smaller events. In 1871 at Gauldry, Fife – a quite small and very rural place –

> Rich and rare plants were kindly sent for exhibition from Birkhill and Mountquhanie gardens, consisting chiefly of stove and greenhouse plants, while Naughton garden was as usual well represented. These occupied the centre table, and were much and deservedly admired. Seldom it is that a provincial show can display such fine specimens of the coleus, caladium, humea elegans, cisus, zonale, begonia, exotic, ferns, etc, as were seen at Gauldry on Saturday.[92]

At Denholm in the Borders, 'as usual, Mr Sutherland, Minto, sent a large collection of stove [greenhouse] plants to add to the attractions of the show'.[93] Not all the specimens sent from mansion gardens were rare or difficult to cultivate. The gardener at Belmont Castle sent twenty-four varieties of gooseberry along with the usual foliage and stove plants to the 1891 Meigle Horticultural and Poultry Show and the Dundee nursery of D. W. Croll sent roses that were probably

88 *Fife Herald*, 18th August 1888, 6.

89 *Dundee Courier*, 21st August 1880, 3.

90 Ibid., 1st September 1880, 4.

91 *Fife Herald*, 2nd September 1880, 8.

92 Ibid., 11th August 1870, 3.

93 *Kelso Chronicle*, 15th September 1871, 2.

Plate 24 Markinch Cottage Gardening Society Annual Show, no date.

put up for sale, as seems to have been the case elsewhere – a chance for rural cottage gardeners to acquire good-quality stock.[94]

The show usually ended in the late afternoon, when the exhibits were removed and the marquee taken down, but the entertainment continued with music and dancing. At Gauldry the show ended with dancing and a concertina contest.[95] At Kennoway in 1877 dancing on the sodden grass beside the recently flooded burn in Kennoway Den went on until late, with the crowded streets illuminated by the lights from the stalls: 'thus a pleasant day came to an end, leaving, it is hoped, good feelings in the community, and a healthy looking forward to another season's emulation in friendly contests of mighty turnips, and tasteful flowers, and gay bouquets.'[96]

Cottage garden shows were much less common in the north-east. Best-kept cottage garden competitions seem to have been more popular amongst cottagers, or perhaps amongst the local landowners, here, while the urban flower shows were the preserve of the middle class. Nevertheless, it was due in large measure to the support from proprietors and farmers, who donated the prizes, that Auldearn, a village in a rich agricultural district near Nairn, was able to hold its first show in 1876. The competitors were mainly 'the possessors of

94 *Dundee Courier*, 10th August 1891, 3.
95 *Fife Herald*, 10th August 1871, 3.
96 Ibid., 16th August 1877, 3.

very small patches of garden ground ... [with] only very little time to devote to horticultural pursuits'.[97]

Flower shows became popular in mining villages in the later decades of the century, where it was almost always the mine owner who was responsible for starting the society. Mr Henderson of Fordell in Fife was a keen gardener and his gardens at Fordell Castle much admired. In 1859 he organised a best-kept garden competition for his miners. The following year Fordell's new society, Mossgreen and District Horticultural Society, held its first show, organised by the colliery manager.[98] In the Lothian coalfields the autocratic coal owner Archibald Hood organised the Rosewell shows. Addiewell, Roslin, Newtongrange and Tranent all boasted horticultural societies. The Newtongrange 20th Annual Flower Show and Industrial Exhibition of 1895 was opened by the coal owner, the marquess of Lothian, himself. As well as flowers, including asters and marigolds, 'great interest was taken in the leek show, and here great perfection was obtained.'[99]

The appeal of the flower show to miners was obvious:

> Horticulture is a favourite pursuit with the men who own the houses they live in, and the Annual Flower Show (which should rather be called a vegetable one) is one of the red-letter days of the year. Very often on the same day as the Flower Show the athletic sports take place, and then the mining village is in its highest gala mood, and it is astonishing how much fun and enjoyment the inhabitants manage to take out of what to an outsider seems anything but a source of happiness. But then if you had grown the biggest cabbage, and your son had won the mile race, the whole complexion of the affair would look differently in your eyes. But look at it as you like, you are unable to deny that the fun and the enjoyment are genuine and hearty.[100]

Cottage garden societies came to be run by the same kind of villager (and often the same villagers) who took most of the prizes. Mining village shows, and sometimes those in smaller villages, were more likely to be dominated by middle-class professionals. On the Mossgreen and District committee sat coal miners and other colliery workers alongside skilled tradesmen such as the blacksmith and the joiner, as well as the School Board (truancy) Officer and a builder – but the show itself was run by the colliery manager.[101] The president and secretary of

---

97 *Aberdeen Journal*, 16th August 1881, 7.

98 Holman, *Behind the Diamond Panes*, 45.

99 *Dalkeith Advertiser*, 29th August 1895, 6.

100 Robert Haddow, 'The nineteenth century' (1888), available at http://www.scottishmining.co.uk/424.html.

101 NRS, GD172/2847/3, Schedule, Mossgreen and District Floral and Horticultural Society, 11th September 1880.

the society at Denholm in the Borders in 1849 were both ministers.[102] It was the parish schoolmaster and a professional gardener who, with the support of two ladies, started the short-lived society in tiny Abbey St Bathans.[103]

Patrons – local landowners, ministers and other rural notables – were essential for the survival of the village show, even one as successful as Markinch. Patrons donated prizes, opened the show, provided grounds and displays of their hothouse plants, lent respectability and head gardeners and – perhaps most important of all – gave encouragement. Societies were correspondingly grateful, welcoming them as honoured guests and presenting gifts. One Borders society boasted no fewer than thirteen patronesses and ten patrons, with addresses ranging from the castle to the bank.[104] Newspaper reports frequently applaud the encouragement and the prizes given to the cottage garden societies by these personages and faithfully recorded their attendance at the shows. The 1861 Ancrum show report chided the gentry for their absence not because it showed disdain or lack of respect for their tenants but because their morale-boosting role was felt to be crucial:

> As this was by far the best show the Society has ever had, it would have been a source of pride and pleasure to have had a good turn out of visitors from amongst the nobility of the district, but, unfortunately, such was not the case. However strong the Society's gratitude may be for their liberal donations and contributions, still their personal inspection and approval is the main stimulant to its future exertions'.[105]

How societies were funded – the contributions of patrons, the sums raised from subscriptions, tickets and entry fees – can reveal something about the character of the societies. The income and expenditure of Markinch Cottage Gardening Society for 1885 was proudly published in the *Fife Herald*.[106] Income amounted to an impressive £148 3s. 2½d, £17 of which had been donated by patrons, while nearly £90 came from tickets and entry fees. The Mossgreen and District show of 1880 in Fordell received £3 2s. 6d from fifty-five very small cash donations, the list headed by the mine owner, G. W. M. Henderson, who contributed £1.[107] There was enormous variety in subscriptions, even amongst thoroughly working-class societies. Kilmadock and Kincardine in Menteith was charging 1s. in annual subscriptions in 1849,[108] but Mossgreen (Fordell) only 6d

[102] *Kelso Chronicle*, 27th July 1849, 6.

[103] *Berwick Advertiser*, 23rd August 1878, 2.

[104] *Berwickshire News*, 5th September 1911, 6.

[105] *Kelso Chronicle*, 20th September 1861, 4.

[106] *Fife Herald*, 16th September 1885, 6.

[107] NRS, GD172/2847/3, Schedule, Mossgreen and District Floral and Horticultural Society, 11th September 1880.

[108] *NBA*, 10th October 1849, 11.

thirty years later. A society in Galashiels for amateurs and cottagers offered free membership to the latter.[109] Mains and Strathmartine Horticultural and Poultry Society in the 1880s distinguished between employer members, who paid 3s., and workmen, whose subscription was half that. This society also ran two best-kept garden competitions, one for each class of member, with the prizes correspondingly graded.[110]

Markinch's expenditure was lavish: £12 for the Fife Marquee, which the Society owned in partnership with other societies, £50 for prizes, £28 10s. 4d for the two bands, over £8 for dinners and refreshment for the bands and the four judges – who had to share a rather measly £2.[111] Societies on a smaller scale, using school premises for their shows and the services of the nearest big houses, would have had much lower expenses.

One of the advantages of the patronage of local landowners was that they could send their head gardeners to judge the exhibits, probably for free. There were other judges and the miner Willie Cook of Donibristle, who was one such, was clearly exceptional and accepted no fee. Bob Holman gives us a striking description of this energetic man:

> A man one could not easily forget – very tall, about six foot, but he was best known for his long red beard. An authority on everything that pertained to a garden, he was well known in the whole of West Fife. A miner by occupation, he had no thought or time for anything else after his work but gardening, and in nearly every flower show in West Fife, and beyond it, Willie was to be seen either judging or giving advice – always free. His own garden in Donibristle was so large that it surprised one to know how he could possibly find time to do anything else in his spare time.[112]

Prizes were donated not only by patrons but by nurseries and seed merchants and local businesses. These could be in kind, although cash prizes were most usual (very often 2s. 6d as first prize). The useful or ornamental household goods and gardening tools presented as prizes give us an intriguing view of the kind of goods working-class families aspired to. Kilmadock and Kincardine in Menteith Cottage Garden Society's show prizes in 1848 included a garden knife, a pen knife, a watering pan, a draw hoe and a flower spade, as well as silver teaspoons, a kailpot, a poker, a coal axe and a looking glass.[113] Mossgreen received donations of prizes from shopkeepers, including two gentlemen's scarves, two pairs of braces, easy chairs, paraffin lamps, linen, a wash hand basin, a clothes line and

---

109 Ibid., 31st January 1849, 12.

110 Dundee City Library Lamb Collection, Mains and Strathmartine Horticultural and Poultry Society Prize Schedule for Competition, 20th August 1887.

111 *Fife Herald*, 16th September 1885, 6.

112 Bob Holman, *Behind the Diamond Panes*, 44.

113 *NBA*, 10th October 1849, 11.

a watering pan. At another mining village show in Midlothian, 'to stimulate the cultivation of plants among the miners and underground workers in the district, substantial prizes in gunpowder were offered by Messrs Hay, Merricks, and Co'.[114]

If accounts of cottage garden societies are rare, still more elusive are records of the activities of their committees. The minutes of the Killearn Cottagers Horticultural Society from 1870 to 1883[115] contain some unexpected dramas. The society was founded in 1850 in the village of Killearn, north of Glasgow. Whilst not part of the Eastern Lowlands, its Lowland village character, as well as the intrinsic interest of the minutes, and their rarity, make a compelling case for citing them. It appears to have been a very successful society (as it still is), with eleven subscription districts and sixteen committee members responsible for collecting the subscriptions. In the minutes there are several entries such as this for 17th January 1874:

> The Committee then entered upon a general conversation on ... the propriety of having a prize for leeks sown and grown in the open garden; on the best manner of exhibiting herbaceous flowers; and of what the basket of vegetables should consist.

No decisions are recorded, however. In 1878 the society was split apart when the professional gardeners walked out, demanding their subscriptions back and refusing to hand over the subscriptions they had collected.

The following year brought an even greater scandal. At the post-show meeting where prize money was paid out, complaints were received from committee members against two members who had entered spring onions and pansies in the show 'contrary to the regulations for the exhibition'. One of the men, Daniel Logan, came forward to claim his prize money in the name of his father, Thomas. The following conversation took place:

John Wilson [the complaining committee member] But your father's dead.

D. Logan	Yes, but the firm is still Thos. Logan and Sons.
The Secretary	We have nothing to do with Thos. Logan and Sons, the stuff's entered in the name of Thomas Logan.
D. Logan	We can easily add 'and Sons' if required.
Secretary	We'll take very good care that you won't do anything of the kind.
D. Logan	Do you refuse to pay?

114 *The Scotsman*, 29th August 1892, 11.

115 I am very grateful to the Killearn Cottagers Horticultural Society for permission to reproduce extracts from these Minutes.

Secretary	No, we wish to delay payment till Saturday first in order you may satisfy the Committee concerning the articles in dispute.
D. Logan	Why did the visiting Committee not examine the garden before the day of the show?
Secretary	The Committee had no complaints about your Garden. But there is a way of settling the matter by which you can sustain your own honour and satisfy the Committee. Just you get two respectable parties to satisfy the Committee that the articles shown by Thomas Logan were really his property and in his possession for six weeks previous to the day of the show.
D. Logan	I will say no more about it tonight.

And with that Logan left the meeting. Three days later he sent a letter to the Society's secretary threatening legal action if his prize money was not paid and a few days after that a Sheriff Court summons was served on him. The case did go to the Sheriff Court but was settled when Logan, evidently a short-tempered and impulsive individual, suddenly agreed to abandon his claim for the prize money.[116]

The anecdote demonstrates the importance of the flower show in village life and gives a glimpse of the social status and social anxieties of its leading members. These small businessmen, prepared to go to court over spring onions and pansies, were clearly familiar with legal proceedings and able to afford them, and very concerned for their reputations.

Society members also relaxed together. Following the autumn show, Markinch's Cottage Garden Society held their annual supper at the Galloway Inn in the town. In 1878 forty members attended, to be entertained with a long succession of toasts to the Society, the patroness, Lady Georgina Balfour of Balbirnie, and each other, and then by humorous, patriotic and sentimental songs sung by members of the company – 'the Queen's welcome to Scotland', 'Riding on the Tramway', 'Tom Bowling', 'Paddy Whack' and 'Rantin 'Robin'. The president was the colliery manager and other committee members were the gasworks manager, a baker, a watchmaker and at least one member of the local volunteer regiment – most of them also winners in the most prestigious classes at the show of 1880, and earlier – and clearly a ruling elite. One of the revellers was Thomas Nicol, who had been winning prizes for flowers at the shows for nearly forty years. In 1880 he took a total of fifteen prizes for potted plants, the older cottage garden favourites like hollyhocks, phlox and snapdragons, and a collection of vegetables. At the supper he did not join in the singing, but toasted and was toasted by the company.[117]

116 Killearn Cottagers' Horticultural Society Records, Minutes, 1870–83.

117 *Fife Herald*, 26th September 1878, 6.

Thomas Nicol, a long-standing member of the parochial board (the precursor of the parish council) as well as the Cottage Garden Committee, appears the archetype of the respectable, public-minded Victorian small town notable. But in some ways he was anything but typical. The son of a Markinch linen weaver, in 1880 he was fifty-five and living in a two-room cottage with his wife, three adult children and a grandchild in the mining village of Milton of Balgonie. His tiny cottage had a half-acre garden where he grew his prize-winning vegetables and flowers. He specialised in enormous onions (one exhibit was fifteen inches in diameter) as well as flowers of all sorts, from larkspur to seven kinds of gladioli. The garden also seems to have accommodated the cows and pigs Nicol regularly advertised for sale in the *Fife Herald*. Another source of income came from his work as a newspaper correspondent (as he described himself in the Census), or 'well-known and indefatigable local correspondent', according to the editor of the *Fife Herald*,[118] who published many short items on natural phenomena and quite a few of his poems. He also contributed several items on Fife gardens, the weather and remarkable plants to the *Journal of Horticulture*. Nicol seems to have been an exemplar, perhaps on a minor scale, of the nineteenth-century self-taught intellectual of very humble origins, who achieved success in his chosen field simply by way of talent and great effort, like many others in this book.

The Markinch show survived until 1959. Kelso was founded in 1827, when many such town horticultural societies began, and is still going strong, but most of the societies have long gone. In the nineteenth century the death, but also the resurrection, of societies was not unusual, although the random nature of the press coverage, outwith Fife, hinders the tracing of a continuous history of any single show. Currie's first show opened with a fanfare but closed down after a few years. It was revived in 1888 and once again was held in the village school. No band is mentioned in *The Scotsman*'s report but butter, oatcake and scone competitions had been added.[119] Many others appear in the press reports for a few years, then disappear. Then, as now, it was the work of energetic activists that kept societies going and fortunate was the village that could maintain the supply.

After 1850 it seems to have become more and more usual for town and county shows to run cottager classes alongside gardeners' and amateurs'. Dalkeith and the Upper District of Midlothian Horticultural Society shows (the latter held in the old conservatory of Arniston House) both had cottager classes, and the Arniston House show even inaugurated a ploughmen's class in 1874.[120] At the same time the mining village shows in Midlothian and many Fife cottage garden shows added gardeners' and amateurs' classes. These developments were probably attempts to broaden the appeal of the shows to both entrants and visitors. This apparent blurring of social class divisions resulted in the successes

---

118 Ibid., 14th December 1887, 5.

119 *The Scotsman*, 27th August 1888, 11.

120 *Dalkeith Advertiser*, 3rd September 1874, 4.

of William Silver and probably many others. Silver was a granite sett maker living in the industrial village of Auchmull (now Bucksburn and a suburb of Aberdeen) alongside quarriers and workers at the paper mill. From at least 1877, when he was in his mid-forties, until 1904, he was winning first prizes in both amateur and working classes categories at the Royal Aberdeen Horticultural Society with his pot plants and his potatoes, kale and celery, as well as taking part in his village show. Entry in these amateur classes was 'open also to the Working Classes on payment of the annual subscription of 5s'.[121]

By the turn of the century it seems the heyday of the cottage garden show was over. In 1871 *The Scotsman* had reported on or advertised fifty-eight late summer shows, of which forty-four took place in the east of Scotland. In 1888 131 shows were mentioned in the six weeks from the beginning of August to the middle of September, but in 1901 only a dozen east of Scotland shows made it into its pages. Was the agricultural depression and its accompanying exodus from the countryside from the 1870s responsible for this decline? Did shows disappear from small villages but flourish in small towns? Was the almost frenetic accretion of additional competitions, from baking to pigeons and sport, a sign of the increasing popularity of the flower show or a desperate attempt to keep them afloat? Or did editors' interest wane? The evidence available at present cannot answer these questions.

THE INFLUENCE OF THE SHOWS

How shows and competitions affected the practice of cottage gardeners and improved their horticulture is hard to judge. Newspaper reports repeat ad nauseam variations on the theme of improvement since the previous year, but as most were probably written by a representative of the society such encomiums need to be taken with a pinch of salt. There are very few other assessments. A few *NSA* reports mention their beneficial influence. The Kilmadock society had been 'productive of much improvement in the appearance of the cottages and of their flower plots and gardens'[122] and the New Pitsligo Horticultural Society 'has excited a considerable emulation among the villagers and tenantry; and the beneficial effects are attested by the superior mode in which their gardens are now kept'.[123] We can point, too, to the rapid growth of cottage garden shows and cottager classes as an indication of the growing enthusiasm for gardening, especially after 1850. On the other hand, this enthusiasm may have more in common with the current popularity of television gardening shows: we don't know how many of the thousands who paid their sixpences to get into the shows went home to trench their vegetable beds or train a fruit tree.

---

121 *Aberdeen Journal*, 22nd September 1877, 3.

122 *NSA* Kincardine in Monteith, Perthshire, vol. X, 1280.

123 *NSA* Tyrie, Aberdeenshire, vol. XII, 724.

The moral message of cottage gardening was never far away, especially in the thoughts of the well-to-do. Many of his contemporaries would have approved of the sentiments of John Philip, chief magistrate of Bonnyrigg, speaking, after several toasts, at the post-show dinner of the Lasswade and Cockpen Horticultural Show: 'He looked upon the cultivation of horticulture by the cottagers as next to religion itself in the improvement of their morality, and as it was for this purpose principally that the society was instituted, he would like to see more and more of them taking an interest in its welfare.'[124]

---

[124] *Dalkeith Advertiser*, 6th September 1877, 4.

CHAPTER 8

THE COTTAGE GARDENER'S EDUCATION

Joseph Banks, director of Kew Gardens under George III and perhaps the best-known botanist of his generation, described Scotland as a 'nation of gardeners', claiming this had been brought about thanks to the 'serious mind of a Scotch education'. Scots professional gardeners and plant collectors ran botanic gardens, gardens of great mansions in England and specialist London nurseries, to say nothing of their European reputation and Empire-wide renown.[1] Yet Scottish cottage gardens and their gardeners were, as we have seen, a cause of some anxiety and despair amongst horticulturalists, social reformers and landowners in the nineteenth century, while comparisons with English cottage gardens were usually to the detriment of their Scottish counterparts.

Throughout our period cottage gardeners were subjected to surveillance, persuasion, guidance, cash inducements, exhortations and occasionally threats in attempts to get them to change their ways. It is impossible to believe that these had no effect. But while not all those with a cottage garden wanted to pay more attention to it than was strictly necessary for raising kale and potatoes, for those who did there were ever more opportunities in the nineteenth century to learn how to prepare the soil, when to sow, the most reliable crops, the hardiest roses and the newest trends. Neighbourly assistance, training, gardening magazines and manuals, school lessons and the educational projects initiated by workers themselves all contributed to those learning opportunities.

In many places it would have been difficult to avoid a never-ending flow of gardening advice and example. The village flower shows, as we have seen, became gala occasions drawing in even the most reluctant gardeners with sideshows, sports, bands and dancing. For those eager to examine the latest and most exotic blooms nurserymen brought their new and exotic plants to flower shows and head gardeners displayed exhibits from their glasshouses.

The part played by conscientious landowners in cottage garden improvement was not limited to the estate competitions described in the last chapter. Many

1 Ron McEwen, 'The northern lads: the migration of Scottish gardeners with especial reference to the Royal Botanic Garden, Kew', *Sibbaldia: The Journal of Botanic Garden Horticulture*, 11, 2013, 109–23, 110–11.

landowners, via their head gardeners of course, supplied cuttings, roots, seeds and 'hints' to villagers. In the early decades of the nineteenth century commentators stressed the importance for cottage gardeners of encouragement by the well-to-do, from landowners to clergy. In fact, without such help, cottagers were not expected to make much progress. HASS and other organisations, as well as horticulturists, never ceased reminding landowners of their responsibilities to cottagers' gardens. Robert Adamson, himself a head gardener, made a point, in his cottage gardening manual of 1851, of lamenting the indifference of most landowners and also called on factors, head gardeners, doctors and ministers to advise and encourage the cottagers, chiding the clergy for frequently neglecting their own gardens and glebes.[2]

Nevertheless, 'cottage gardeners were the last to be edified by sound teaching', declared the gardening historian Edward Hyams.[3] Hyams' argument was that new varieties 'trickled down' to cottage gardens only several decades after their introduction, and that old, often nonsensical, practices survived there long after professional gardeners and experts had abandoned them. While this was undoubtedly true of many, perhaps most, there is at the same time plenty of evidence that there were cottage gardeners who wanted to learn, who found opportunities to acquire new knowledge and who, as the last chapter made clear, carried off prizes in proof of their success. Some took the pathway from gardening into botany and the scientific study of all aspects of nature. The Victorian fascination with botany drew in some notable working-class gardeners and a few achieved national and even international fame.

CHILD GARDENERS

Boys (and probably girls too) learned to garden through many routes, very often through their own wits and determination, although helpful and generous neighbours are sometimes hovering in the background. Trial and error, with a lot of the latter, was the way the Bethune brothers learned to garden as boys – there is no mention of their parents teaching them. Alexander recounted how, after a remarkably long time, they finally realised that it was lack of watering that was killing off their young gooseberry bushes, raised from slips provided by their neighbours, until 'at last, observation and experience, aided by a long succession of failures, enabled us to succeed in raising young bushes.'[4]

The Borders poet and shoemaker John Younger made his first garden at the age of thirteen, unaided it seems, and he boasts in his autobiography of his instant success with cabbages. Younger was the son of a small farmer and grandson of a trained gardener who – according to Younger – could have been

2 Adamson, *The Cottage Garden*, 12.

3 Hyams, *English Cottage Gardens*, 69.

4 Bethune, *Poems by the Late John Bethune*, 72–3.

another Dickson of Hassendeanburn (see below) had he not lost all his money in a failed brewery venture.[5] Younger's gardening trials also benefited from the laird, who rode by his garden and was so taken with this efforts that he told him how to build a stronger fence to keep out the neighbours' hens and gave him permission to cut enough broom and briers to make it with.[6]

Another shoemaker, the naturalist Thomas Edward of Banff, also learned early on from his mistakes in the garden. As as a boy in Aberdeen he tried his hand at botanical gardening, clearing a small rubble-strewn area behind his house, preparing the soil and planting wild flowers he had collected from the woods, fields and the banks of the Dee and the Don. In spite of watering and tending them, and replacing those that died, his 'wild botanical garden' was a failure. Raspberries, wild strawberries, foxgloves and some ferns and grasses survived, but all the other flowers perished.[7] Samuel Smiles' biography mentions the manure Edward applied to his garden, and this was probably what did for the meadow flowers. The same fate overtook his collection of frogs, tadpoles and newts. On his mother's advice, Edward gave up his wild garden and created a more conventional and successful plot, with plants rescued from various dumps:

> The neighbours, who had first sneered at him as a fool, on seeing his pretty garden, began to whisper that the 'loon' was surely a genius and that it was a pity his father had not made him a gardener instead of a shoemaker. Edward himself often wished that his parents had been of the same mind as the neighbours.[8]

These boys, teaching themselves by trial and error, were probably exceptional – certainly they grew up to be exceptional men. Labouring in the gardens and estates of middle- and upper-class families is likely to have been a more common way for those who ended up with cottage gardens to learn about gardening, if they did. Many would have been boys, perhaps not all as reluctant as Nathaniel Paterson's. Paterson's 1836 book on manse gardening, based on his own practical experience in Berwickshire, ends with an Appendix on the Minister's most useful garden implement – a boy – and how to train and keep him. This boy would have been about fourteen and 'somewhere between an urchin and a varlet', according to Paterson; rebellious, not brought up to habits of industry, longing for the busy, sociable and eventful life of the farm or the idle gossip of the smithy, but condemned to a solitary and heavily supervised life of drudgery in the minister's household. Paterson's advice, often tongue-in-cheek, is for the

---

5 Younger, *Autobiography*, 12, 14.

6 Ibid., 106.

7 Samuel Smiles, *The Life of a Scotch Naturalist*, London 1905, 77.

8 Ibid., 78.

minister himself to train up this lad in gardening, and in morals and education as well. He will double dig, rake and weed, but expect him to leave for a farm job as soon as he can, warned Paterson.[9]

WORKING IN MANSION GARDENS AND NURSERIES

Paterson's lad may have found himself working in gardens again when farm work was not on offer. Labourers, obliged to seek work wherever they could, might work in nurseries, as jobbing gardeners in middle-class homes or in a farmer's garden. There were also armies of under-gardeners, journeymen gardeners, male and female labourers and boys employed in the gardens and plantations of the new mansions built by the rising upper and middle classes. The laying out of policies and creation of designed landscapes required a vast amount of earth-moving, planting of trees and shrubs, building of walls and digging out of ornamental lakes. The erection of walled kitchen gardens (often enormous – Blair Castle's walled garden fills nine acres), glasshouses, hot beds, summer houses, arbours – all these projects and tasks demanded heavy labour. Once the garden had been created weeding and other lighter unskilled maintenance tasks gave work to women and boys.

A contemporary statistical analysis of the 1841 Census calculated that there were 170,000 'labourers' in Scotland, of whom around three per cent (6,418) were gardeners, a smaller proportion than in England and Wales.[10] Although these figures may be questionable, not least because the term 'gardener' was not clearly defined, they at least give an idea of the scale of gardening employment.

Patrick Neill observed that, where cottagers had worked as labourers in landowners' gardens, a few of their gardens could be found in good order, producing a variety of vegetables, although he does not elaborate, unfortunately.[11] We know about some of the tasks the labourers carried out from estate papers. At the very grand estate of the Clerks of Penicuik, not far from Edinburgh, between 1817 and 1826, up to twenty-two workers were employed at day rates in addition to the gardeners hired by the year. They cut weeds, repaired fences and hoed the garden and the shrubbery. In the early summer women were hired to weed, and in early winter they lifted carrots. In the early spring of 1820 the entire workforce of twenty-one, which included four women and three boys, turned out to plant trees, so even the lowliest workers, presumably supervised by the professional gardeners, gained a variety

9 Paterson, *Manse Garden*, 246 passim.

10 Anon., 'Our agricultural labourers', *Journal of Agriculture*, July 1849–March 1851, 440–8, 441.

11 Neill, *Gardens and Orchards*, 22.

of gardening experience.[12] Much later in the century, a more modest garden at Fordell Castle, in Fife, had a staff of seventeen in 1879, with four journeymen gardeners, seven labourers, two boys and a 'garden woman', for whom the total wage-bill amounted to well over £600.[13]

Nurseries thrived in the early decades of the nineteenth century. The Post Office Directory for Edinburgh for 1820–1 lists at least eight nurseries, and there were six ten years later. The Aberdeen Directory for 1830–1 gives seven.[14] Later, numbers of provincial nurseries fell, perhaps, as E. H. M. Cox asserts, because the planting fever was abating.[15] Nurseries needed labourers as well as skilled gardeners, especially during the improvement era, when mansions were being put up at speed and newly laid-out grounds demanded great numbers of trees and shrubs. The most extensive and successful nursery business in Scotland for nearly two centuries was run by the Dickson family. After starting their first venture at Hassendean in Roxburghshire in 1729, they opened a second establishment in nearby Hawick in 1766, employing, by the 1790s, up to fifty workers on thirty-six acres on the outskirts of the town.[16] A year later, James Dickson opened a nursery on the slopes of Kinnoull hill, just outside Perth, where the firm also had a shop. At Kinnoull in the 1790s they had around forty acres and a workforce of forty. Forty years later, this nursery covered sixty acres and employed between seventy and eighty workers.[17] The Dickson nurseries, owned by various branches of the family in the Borders, Perth and Edinburgh, supplied gardens and estates throughout Scotland, the north of England and Ireland with trees, fruit trees, plants and shrubs of all kinds. Much smaller places contained nurseries too, often considerable. Nurseries in Banff, Crieff in Perthshire and the village of Chirnside in Berwickshire all appear in the *OSA*. Mr Reid's nursery near Banff, opened in the 1760s, covered fifteen to twenty acres and grew fruit and forest trees, evergreens, shrubs, 'roots' and plants.[18] This nursery and the one at Crieff were still flourishing in the 1830s.

Alexander Somerville worked as a young man for eight months in the Edinburgh nurseries of James Dickson and Son at Inverleith, where J. C. Loudon had been apprenticed. His fellow workers were a mixture of well-educated, middle-class young men who had studied botany, labourers like himself and experienced professional gardeners out of a place and 'submitting to work for

---

[12] NRS, GD18/1735/2, Day labour books, Clerk of Penicuik estate papers, 24/11/1817–25/12/1819.

[13] NRS, GD172/764/2, Memorandum on arrangements to take effect at Fordell at Martinmas 1879, Henderson of Fordell papers.

[14] Scottish Post Office Directories, NLS e-resources, available at https://digital.nls.uk/directories.

[15] Cox, *Gardening in Scotland*, 170.

[16] *OSA* Hawick, Selkirkshire, vol. VIII, 530.

[17] *NSA* Kinnoull, Perthshire, vol. X, 941–2.

[18] *OSA* Banff, Banffshire, vol. XX, 331.

9s. per week in the hope of getting other situations as masters'. Somerville's tasks were to trench and dig shrub and tree beds and barrow manure, although he also had some (unspecified) training.[19]

As an unschooled and unskilled teenager, Alexander Bethune's brother John was employed as a labourer at the new mansion of Inchrye in north-west Fife, the largest and grandest of four new mansions in the parish of Abdie, built at a cost of £12,000 in the Gothic style, complete with turrets and battlements. He quickly learned on the job and became – briefly – an expert and valued head gardener. His career illustrates the opportunities for social mobility that this world presented, but also how easily they could be snatched away. In spite of never having been apprenticed and having no formal education – he spent no more than one day at school – he worked there entirely unsupervised for seven years.[20] He learned his trade so thoroughly that at the age of nineteen he was appointed head gardener and took on his older brother, Alexander, as assistant. Alexander recounted how thoroughly and creatively his brother set about his new tasks:

> A number of repairs and alterations being required on the estate, the proprietors wished a written report as to the means to be employed, and the most economical method of effecting them; and as he now considered his powers, both of body and mind, engaged in the service of a master, shortly after entering upon his new labours he gave up literature entirely and devoted the whole of his time during the evenings [in the winter months] to this single object. By inventing a very simple instrument for taking angles, and another for laying them upon paper, plans of the various fields and plantations were drawn, with what appeared to be the necessary alterations marked upon them. An estimate of the probable expense to repairing the hothouses, together with the means by which it could be done, was made out; and as a proof of the correctness of his ideas, it may be mentioned, that in executing these repairs, on the following year, the very plan which he had pointed out was executed by a proprietor … . In the spring there was a great amount of labour to perform in enclosing grass fields, which were to be let for pasture, while at this season the hot-houses would have required nearly the whole of his attention. That no expense might be incurred which he could possibly save, he undertook a large share of the new fences himself: at these he wrought diligently during the usual working-hours: and that the hot-houses might not be neglected, his mornings and evenings, as early and as late as daylight would serve, were devoted to arranging the clusters, and tying down the young shoots of the vines – his father attending to the temperature while he was absent. With the rigid attention which he bestowed on them, things seemed to prosper: the enclosing of the fields had been completed in good time, a luxuriant crop of grapes was advancing to maturity.[21]

19 A. Somerville, *Autobiography*, 99.

20 Bethune, *Poems by the late John Bethune*, 118.

21 Ibid., 61–2.

At this point, a new proprietor arrived and let John go. He went back to working as a road-mender.

John Bethune was only an 'acting' head gardener, but was paid £26 a year, considerably more than a ploughman at the time. Qualified head gardeners could command high salaries and were highly prized, and sometimes feared, by their employers: the 'Scotch' head gardener, independent, knowledgeable, highly skilled, argumentative to the point of cussedness, was a familiar figure from the eighteenth century on, long before he made his appearance in the novels of P. G. Wodehouse. Head gardeners were at the apex both of the staff of a great house and of the rural working class, men of power and influence within the noble household and the surrounding countryside (witness their role at flower shows). They were often sons of gardeners or upper estate workers and well educated at the parish school. Most aspirant gardeners, not all of whom achieved headship, went through tough apprenticeships, followed by 'internships' of a year at different gardens where they honed their skills.

Evidence recently unearthed by Fiona Davison at the Royal Horticultural Society has disclosed the backgrounds of a chosen few trainee professional gardeners. These young men, recruited from all over Great Britain, were selected for a science-based training course by the Horticultural Society, founded in London in 1804 by a group of botanists, gardeners and enthusiastic aristocrats. These trainees included Joseph Paxton and went on, as Fiona Davison has recorded, to create, run and work in great gardens in Britain and across the empire.[22] The Horticultural Society's 'Handwriting of Under Gardeners and Labourers' book[23] contains the CVs of 105 trainees taken on between 1823 and 1829.[24] A significant, and disproportionate, number – twenty-nine, just under one third – came from Scotland, at a time when the population of Scotland was only one sixth that of England and Wales. Of these no fewer than eight came from Perthshire. Most were, like the English men, the sons of gardeners, small farmers and upper estate workers such as foresters; others were sons of small businessmen or tradesmen. None had, like John Bethune, labouring or even farm servant backgrounds. The Horticultural Society asked for a recommendation from a Fellow of the Society, which implies good social connections, and previous gardening experience was required as well as literacy. The 'Handwriting' book reveals that the Scots were better educated than the English, with on average two more years of schooling, having left school at fourteen. These young men were or became the cream of the gardening profession, but most professional or working gardeners would have had a much less sophisticated training and never

[22] Davison, *The Hidden Horticulturists*.

[23] RHS Lindley Collections, RHS/VS/hist/10.2, 'The Handwriting of Under Gardeners and Labourers'.

[24] I am grateful to Fiona Davison for providing biographical details of the young men drawn from the Handwriting book.

have risen to more than journeyman status. As such, they worked long, arduous hours, living in the gardeners' bothy and suffering the autocracy of the head gardener even more than their employer.

GARDENING MAGAZINES AND GARDENING MANUALS

Professional gardeners had been well supplied with manuals and journals since the publication of John Reid's *The Scots Gard'ner* in 1683. Their employers bought them, and allowed their gardeners into their libraries to read them. J. C. Loudon's *Gardener's Magazine*, launched in 1826, was also aimed at professional gardeners. It 'had two grave objects in view; – to disseminate new and important information on all topics connected with horticulture, and to raise the intellect and the character of those engaged in this art'.[25] The price undoubtedly restricted its readership: 5s. for each quarterly issue, even when reduced to 1s. 6d in 1834, would have been expensive for most head gardeners. But Edward Hyams has argued that

> it is likely that through [working gardeners] the magazine's teaching was passed on to hundreds of cottage gardeners, and sometimes, no doubt, the magazine itself. Be that as it may, it is certain that the *Gardener's Magazine* was the first horticultural periodical to be of use, if only at second hand, to the cottager.[26]

As a social reformer Loudon was keen to promote cottage gardening and, as well as reprinting reports of shows and frequently highlighting the achievements of cottage gardeners, he published regular contributions on the topic, such as, for Scotland, Gorrie's and Menteath's on ways of improving methods and crops.

Would there have been any point in publishing manuals and magazines to be read by cottage gardeners at this time? Literacy in Scotland was higher than in England, as it had been since the Reformation. The parish school system mapped out by John Knox in 1560 succeeded in educating the majority of the male population of the Lowlands until the rapid population growth of the later eighteenth century. It is likely that nearly all men in Scotland could read by 1800, even though by no means all children attended school. Many were too poor or lived too far from the parish school; the schools charged fees, each skill being charged separately so many, and most girls, learned only to read.[27] The most common experience for poor children was a year or two's attendance, from the

[25] *GM*, 1826, Preface.

[26] Hyams, *English Cottage Gardens*, 104–5.

[27] Estimates based on signatures in marriage registers give an adult male literacy rate of eighty-eight per cent around 1800, compared to sixty-five per cent in England, but this is likely to be an underestimate (R. D. Anderson, *Education and the Scottish People 1750–1918*, Oxford 1995, 15–16).

age of seven often not at the parish school but at informal schools taught by old women or young men. And, as they were working throughout the summer, a year meant the winter months.[28]

Literacy, however, never depended only on formal schooling. Working-class memoirs bear witness to the many parents, grandparents, friends, neighbours and impecunious students who taught children and teenagers to read and write, like the Bethune brothers and John Duncan the weaver-botanist. There were books, chapbooks and newspapers to share and read. Most homes had a family bible and perhaps a few collections of sermons and there were cheap pamphlets to buy from the cadgers, but for those with an appetite for reading there were parish and circulating libraries and later the Mechanics Institutes and Mutual Improvement societies. At the beginning of the nineteenth century free or cheap subscription lending libraries – Innerpeffray library near Crieff, the oldest free public lending library in Scotland, Leadhills and Wanlockhead miners' libraries in the south-west and Samuel Brown's 'itinerating library' in East Lothian, and many more – brought books to rural workers or were founded and run by them. Weavers were often the keenest borrowers, as at Innerpeffray, but in the Cheviot hills Border shepherds were said to maintain a kind of circulating library, leaving books in crannies in dykes.[29] Such libraries contained mainly books on religion, popular science, travel and history, but occasionally agriculture and horticulture. According to *NSA* reports there was a library in almost every parish: Sunday school or parish for the most part, but many others. An 1840 catalogue for the village library in Rait in Perthshire is now on display in the village hall. This library was started by Archibald Gorrie and the catalogue includes a copy of Loudon's *Manual of Cottage Gardening and Husbandry* of 1830.

Newspapers and periodicals were running gardening columns from early in the nineteenth century and, although they were out of reach for most workers,[30] newspaper reading groups, where each member contributed towards the price of 8d or 9d, were not unusual, or a copy might be passed on from a better-off neighbour.

The Cheap Magazine; or Poor Man's Fireside Companion, first published in Haddington in 1813, was one of the very first periodicals aimed explicitly at the rural working class.[31] A monthly, it was priced at 4d for fifty small pages. Describing itself as 'A work of Humble Import; yet claiming the attention of all ranks, as having for its object the Prevention of Crimes, ... by alluring the young and thoughtless to a taste for reading subjects of real utility ... especially for the lower orders of mankind', the aim of the editor and publisher, George Miller

[28] Ibid., 3–8.

[29] Jonathan Rose, *The Intellectual Life of the British Working Classes*, 2nd edn, New Haven 2010, 60.

[30] The Stamp Act of 1815 – the 'tax on knowledge' – added 4d per copy to the price of a newspaper until 1836, when it was reduced to 1d. It was not abolished until 1855.

[31] *The Cheap Magazine*, Haddington 1813.

of Dunbar – awarded the title of 'the father of cheap literature' by Alexander Somerville[32] – was to provide a stimulating and entertaining diet of moralising tales and uplifting biographies (The 'Reformed Prostitute' and the life of Caravaggio stand out), poetry, travellers' tales and useful information and advice, written in a lively and unpatronising style.[33] Its low price was never matched by an adequate distribution and it ceased publication after two years.

The *Cheap Magazine* had a monthly garden advice paragraph entitled 'Work to be done in the Cottager's Garden'. It was strictly a calendar, probably reprinted from a gardening publication or an almanac, and did not attempt to explain gardening techniques. Only vegetables of the most common type and soft fruit bushes were dealt with regularly, with passing references to annual flowers. For May, the advice was: 'continue to sow Cauliflower, Savoys, and Cabbages, for a late crop. Sow full crops of French and Turkey Beans – and Pease for a last full crop. Thin Onions, Carrots, Turnips, and Spinage [sic]. Be diligent in clearing off weeds and destroying vermin.'[34] Some of the advice was optimistic. In December cottage gardeners were recommended to sow radishes, peas and beans if the weather was mild and to transplant gooseberries and currants.[35] There was also a regular botany column, which dealt with seeds, buds and winter sowing times.

Chambers Edinburgh Journal, launched in 1832, was the first successful inexpensive popular periodical in Scotland; priced at three halfpence for eight pages, its weekly circulation soon rose to 84,000. Its reach and influence would have been many times that as each issue was shared among several readers. *Chambers*, as it became known, contained news, political comment, articles on travel, science and history, and short stories (the Bethune brothers published some of their stories in it). From the very first issues it ran articles on garden vegetables and fruit. Initially, regular gardening advice was summarised from George Lindley's *Guide to the Orchard and Kitchen Garden* (1831) on such topics as hot beds, grafting, grapevines and cucumber frames. Subsequently this was dropped and articles on topics more relevant to the readers, such as fruit-tree grafting, the history of the potato, renovating old trees, weeds, the productiveness of vegetables and soil structure were dealt with not only technically but also scientifically. In 1835 it ran a 'Column for Cottagers', as, according to its author, 'thousands of cottagers were reading *Chambers*'.[36] This ready-made readership did not deter the anonymous author from launching a scathing attack on them: cottagers allow themselves plenty of time to smoke, drink and talk (politics), but neglect their homes and fail to whitewash the walls. More helpfully, he recommended the digging of drains to direct household waste

32 A. Somerville, *Autobiography*, 60.

33 The 'Useful Information' feature was much more likely to cover how to rescue people from fires, drowning and falling through ice than anything more mundane, if less likely to appeal to the young readership.

34 *The Cheap Magazine*, Haddington 1813, vol. 1, 236.

35 Ibid., 570.

36 *Chambers Edinburgh Journal*, vol. III, 1835, 368.

into gardens for manure and the planting of honeysuckle and sweetbrier on the newly whitewashed walls. He also recommended the perennial flower marvel of Peru (*Mirabilis jalapa*), a Victorian favourite, though somewhat tender for Scotland. The result of all this effort and expense will be that the children will grow up with 'a better taste for the beauties of nature', and will learn to be neat and to respect nature. This may seem a rather meagre, and remote, return and it is worth noting that this author signally fails to call on landowners to assist, except in the matter of the drain-digging. Perhaps this article did not go down well with the readers, as the 'Column for Cottagers' does not seem to have continued.

Cheap gardening magazines, aimed at all classes of gardener, notably Joseph Paxton's *Horticultural Register* and his *Gardeners' Chronicle*, started to appear in the 1830s and 1840s.[37] The *Cottage Gardener*, a 3d weekly, was launched in 1848, with contributions on growing vegetables and hardy plants by head gardeners and nurserymen, but the fastest growing market for gardening magazines was by now amateur gardeners, increasingly the new lower-middle-class suburbanites.[38] Working-class gardeners were not neglected, though, as more and more ever-cheaper weeklies were launched. *Gardening Illustrated* was a new William Robinson publication, begun in 1879, the cheapest ever at 1d and explicitly aimed at 'the smaller class of gardens' with a focus on cottage and window gardening.[39] By the 1890s there were eight national gardening weeklies, all cheap but 'all ... drearily similar' and lacking any originality, according to Anne Wilkinson.[40] In Scotland there was *The Gardener*, for amateurs and professionals, edited by the duke of Buccleuch's gardeners. The fortnightly *North British Agriculturist* (1849–1893) and its sister magazine *The Scottish Gardener* (from 1851) published frequently on cottage gardens and cottage garden shows, but the price (12s. per annum for the former) and the rather condescending and occasionally even hectoring tone suggests they were not designed for a cottager readership: 'it would be well if some of your correspondents would shame occupiers of a farm or cottage garden by referring to [the] slovenly and unprofitable appearance [of their gardens]', wrote one concerned correspondent.[41]

Such a tone seemed to have pervaded much of the gardening literature as it related to the cottage garden at this time. A contributor in 1870 to the Edinburgh- and London-based *The Gardener* hit back at the current middle-class and, in their view, patronising concern for cottage gardeners:

37 Anne Wilkinson, 'The Development of gardening as a leisure activity in 19th century Britain and the establishment of horticultural periodicals', unpublished PhD thesis, Open University 2002, 101.

38 Ibid., 118.

39 Ibid., 133.

40 Ibid., 142.

41 *NBA*, 10th January 1850, 11.

> there are two things about cottagers that seem to have escaped the attention of their would-be patrons – 1st, They take but little interest in garden literature (the horticultural column of a penny weekly paper being their chief reading); 2nd, They are, as a rule, much better versed in practical cottage-gardening than they have hitherto been supposed to be.[42]

The article goes on to argue that the cottager knows the value of good digging and the value of manure, and that these are what are crucial for producing good-quality vegetables. Cottage gardeners find nothing relevant in the 'horticultural organs' and what they need is their own magazine. Cottage garden societies, along with help, advice and improved seed varieties and fruit trees handed out by provincial clergy and landowners, would do more to raise the standards of poor people's gardens than any number of prize essays.[43]

No Scottish horticultural organ aimed specifically at cottagers appeared – or at least has come to light. Columns in the cheap national newspapers such as the *Penny Illustrated* (in the 'Family Column') and gardeners' kalendars in the almanacs found in most working-class homes were probably the most common source of information. George Glenny, a prominent founder of gardening magazines for working people as well as the middle classes, contributed a column to the *Edinburgh Weekly Journal* from 1848. Oliver and Boyd's *Edinburgh Almanac* – the publication posted to all coastal lighthouses by the Northern Lighthouse Board – included a Gardener's Kalendar listing a lengthy and alarming list of monthly jobs including grafting roses and fruit trees and making hotbeds for melons.

Manuals designed for cottage gardeners were even harder to come by. In 1834 *Chambers* recommended a work allegedly written for the cottage gardener, *The Hand-book of Gardening* by James Rennie, a Scots-born professor of zoology. In fact this was a scientific treatise including such arcane instructions as 'in order to understand the nature of the food or nourishment of growing crops, as well as the mode of their feeding, it will be necessary to examine the mouths and other parts of garden plants, such as turnips, gooseberry trees, or lilies.' Loudon published a typically abrasive review in the *Gardener's Magazine*: 'This little book ... deserves to be favourably received by the public, from the excellence of the intentions of the parties connected with it. In our opinion, however, it is by no means the sort of work calculated to attain the end in view.' He went on to contend that no 'mere London author' could achieve this and that 'the man who, of all others that we know, either in Scotland or England, could ... is Mr Gorrie'. Loudon goes on to urge *Chambers* to put out a 2d pamphlet on the subject – all the Scotch cottager wants could be contained in this.[44] So far as is known,

42 Anon., 'Horticulture for cottagers', *The Gardener*, August 1870, 337–41, 337.

43 Ibid., 340.

44 *GM*, 1834, 616.

Archibald Gorrie never produced such a work. Loudon's own book, *A Manual of Cottage Gardening, Husbandry and Architecture* (1830), was no more designed for the cottager than Rennie's, however.[45]

Cottage gardeners in Scotland had to wait until the 1850s, so far as is known, for their own handbook. Robert Adamson, the head gardener at the Balcarres estate in Fife, well known as a flower show prizewinner and judge, as well as a contributor to many mid-century gardening magazines, may have published the first manual of practical use for cottagers in Scotland, although whether many cottage gardeners bought this, or any others, is not known. *The Cottage Garden* (1851) covered every aspect of gardening, from materials for 'walks' to crops and tools and from 'garden cooking' (pickles, jelly, jam and wine) to the destruction of vermin. In the Introductory Remarks he makes it clear that he has 'guarded against recommending anything inconsistent with such limited means, as a cottager is supposed to possess'. Nevertheless, the list of tools for 'anyone not very particular' is daunting: wheelbarrow, spade, dung-fork, fork, draw-hoe, three-pronged draw-hoe, Dutch hoe, large rake, small rake, reel and line, hardwood dibble, pruning knife and watering can![46] A new spade at this time in Fife could cost 3s. 6d, equivalent to two day's wages for a labourer.[47] Adamson is more parsimonious in his choice of vegetables. Those that can be harvested over a long period, that suit the soil and situation and that store well should be preferred, including brussels sprouts and spinach, rhubarb (by now a culinary crop), potato onion and garlic and new varieties of old favourites.[48] Herbs for cooking (balm, hyssop, spearmint, sage and knotted marjoram, amongst others), perfuming (such as sweet basil, rosemary and lavender) and 'stomachic' (horehound, chamomile, tarragon) are recommended.[49] Eight pages are given to lists of flowers – none of them hothouse species – and 'a few of the most useful roses' suitable for cottage gardeners, all well known, such as China, Scotch, damask and sweetbrier. Adamson begins his manual by specifying the need for enclosing the cottage garden with a wall, fence or hedge, to deter 'pilfering bipeds and quadropeds'. The walks or paths, which create subdivisions for fruit and vegetables, should be made up of free and locally available materials. His list of these reveals quite a bit about the physical context of cottage gardening: 'pit, river, or sea gravel, rotten rock, engine, or other furnace ashes, lime riddlings, brick rubbish, coarse pit sand, road drift, tanner's bark, or sawdust'. Walks

45 Tellingly, the *Spectator* review recommended the work 'warmly to noblemen and gentlemen, who are from interest, if not from higher motives, desirous to contribute to the comforts of their labourers' (even though 'Mr. LOUDON has directed the work to be sold without profit, in order to render its general distribution as inexpensive as possible'), *Spectator*, 8th May 1830, 45.

46 Adamson, *The Cottage Garden*, 20.

47 *NSA* Carnock, Fife, vol. IX, 708.

48 Adamson, *The Cottage Garden*, 9–10.

49 Ibid., 16.

should be edged with box or small flowers such as daisies or thyme, or pebbles.[50] He also recommends frequent (successional) sowing and stresses the importance of rotating garden crops.[51]

The next cheap manual, not published until 1889, at 1s., was probably Alexander Sweet's *Villa and Cottage Gardening specially adapted for Scotland, Northern England and Ireland*. This sold well and had gone into eight editions by 1906. He addresses the 'weary artisan' with 'a bit of garden' (less than half an acre, although such a very small garden limits the opportunities for rotations).[52] Reliable varieties of hardy fruit trees – pears, apples and plums – suitable for northern climates are recommended[53] and he lists only the well-established and easy-to-grow flowers and vegetables. Greenhouses are never mentioned. Alexander Sweet was not a professional gardener or a cottager: he worked in a bank all his life, but his very practical and realistic handbook is the result of many years of gardening experience at his house in the Glasgow suburb of Cathcart.

Sweet's recommended layout is traditional, and echoes that of Adamson forty years earlier, paying no homage to the latest fashionable garden designs. The front garden should be laid out in grass and a few ornamental beds – 'far better to be prosaic than so ornamental as to outrage the whole look of a place by being far-fetched and meaningless'.[54] The back garden should be divided into quarters for grass, fruit bushes, flowers and vegetables; the walks once again are to be edged with box or with cheap, compact and neat flowers such as daisies, saxifrage or thrift.[55]

The message to cottage gardeners in Adamson's and Sweet's manuals was the same: the garden was above all to be useful and – inevitably – neat, although flowers, especially in Adamson's book, had their place too. Interestingly, neither of them had anything to say about entering flower shows.

SCHOOL GARDENS

Ideally, a gardener's education should start early. Proposals for including gardening in the parish school curriculum crop up from time to time,[56] starting with James Stuart Menteath's much quoted and far-sighted article published in

50 Ibid., 2.

51 Ibid., 11.

52 Alexander Sweet, *Villa and Cottage Gardening Specially Adapted for Scotland, Northern England and Ireland*, 10th edn, London 1921, 4–5.

53 Sweet's recommendations included the apples James Grieve, St Edmund's Russet, Rosebery, Worcester Pearmain and Ribston Pippin; Crawford and Jargonelle pears (the others needed walls); and the relatively new Victoria and Stint plums (Ibid., 60–2).

54 Ibid., 45.

55 Ibid., 55–6.

56 See D. Gorrie, 'Hints on the improvement of cottages', 1.

the early 1830s.[57] Menteath argued persuasively for the teaching of 'common kitchen gardening' to schoolchildren but there is no evidence that his advice, or that of other writers, was heeded until much later. He proposed that each school have half an acre of land, divided into small plots, each of which would be worked by a group of children; berry bushes could be used to divide the plots and fruit trees planted for grafting and pruning lessons. Teachers could be taught botany at university (parish dominies were required to have attended university) and learn the practical side from the nearest head gardener in the summer holidays. Adding gardening to the curriculum would entail almost no disruption to the life of the school or the work of the teacher – gardening lessons could take place during play time and on Saturday and the schoolmaster's garden, fixed by statute at one rood (one quarter of a Scots acre), would provide at least some of the land. In addition the teacher would receive all the produce and could pay the rent on the rest of the land by selling some of it. The children could be rewarded with prizes of packets of seed, garden tools and books, which would also help their families; neighbouring gardens and orchards would be preserved from juvenile predators; 'the youth of the country will grow up with a love of rural affairs' and physical and moral improvement all round would ensue.[58]

The difficulty of finding schoolmasters with a basic knowledge of botany and of how to grow common vegetables and fruit (and a 'slight acquaintance with flowers'[59]), to say nothing of the extra effort overstretched dominies would have had to put in (and the potential for the wrecking of their gardens) was probably what doomed this and other equally well-meaning proposals.

Nevertheless, there was one unique and very successful venture in incorporating gardening into education. Dollar Academy in Clackmannanshire was founded in 1818 with an endowment from a local merchant for the education of the poor children of the village of Dollar. The Academy was encircled, and still is, by extensive grounds planted with ornamental shrubs and trees. Vegetables were grown and there was a commercial nursery. Gardening was taught as a healthful and useful out-of-school activity for its pupils and at the same time there was training for boys hoping to become professional gardeners. The *NSA* report for the parish of Dollar, written by the Academy's founding rector, Andrew Mylne, describes how a number of the pupils, on completing their general education at the school, were

> allowed to enter the gardens as apprentices, where, under the supervision of an experienced and scientific gardener, they are instructed in all the branches of gardening. They also attend a class for drawing in the evening; and with

57 Menteath, 'On improving cottage gardening in Scotland', *Quarterly Journal of Agriculture* 1831–2, vol. IV, 783–6.

58 Ibid., 785.

59 Ibid., 783.

> some other pupils, both male and female, receive lessons in botany from the master gardener. In this way, they become qualified for places in gentlemen's gardens.[60]

The Dollar apprentices went on to take charge of gardens here and across the world – in India, Constantinople and New Zealand.[61]

The master gardener was John Westwood, a former pupil of the school, appointed in 1838. It was John Westwood who made the Academy gardens famous. A popular and charismatic teacher, he taught gardening and botany and trained the apprentices as well as taking groups of pupils on nature walks on Saturday afternoons. After his death in 1875 the apprenticeship scheme lapsed and sports pitches were laid out on the flower and vegetable beds.[62]

As well as this professional training, however, eleven of the 'best behaved and most diligent pupils' tended small plots in the garden and could take their produce home, according to a memoir by George Lawson, a pupil in the 1850s.[63] They could work their plots for four years from the age of eleven, and two prizes were awarded each year for the best-kept, with other prizes given at the Annual Exhibition, 'now and again'. A boy who failed to keep the plot tidy, however, would have it taken away.

Mid-Victorian city-based industrial schools for poor children taught gardening along with other trades, but the idea that gardening should or could be taught formally to children in rural parish schools and not as part of an apprenticeship was foreign to most thinking of the time. Science subjects had almost no place in elementary or high school curricula and gardening was seen as the kind of skill that should be passed from parents to children – in fact it was the responsibility of the labourer or farm servant to instruct his children in the garden as part of their preparation for agricultural work. There were odd exceptions. In the Berwickshire fishing village of Eyemouth there had been school gardens since at least the 1850s and in 1875 the *Berwick Advertiser* carried a letter from the schoolmaster reporting on the successes and failures of that year. He set a kailyard syllabus: the boys grew leeks, cabbages, potatoes, onions and kale. Apart from one well-managed plot, their efforts in the summer of 1875 were rated as poor ('leeks require weeding'; 'still maintains its bad character'), mainly because most of them, as fishermen's sons, spent their summers at sea on the herring boats.[64]

---

60 *NSA* Dollar, Clackmannanshire, vol. VIII, 99.

61 *Dollar Magazine*, vol. V, September 1906, 19.

62 I am grateful to Mrs J. Carolan, Dollar Academy Archivist, for the information about John Westwood.

63 George Lawson, *Reminiscences of Dollar Academy, Church and Sunday School*, 3rd edn, Alloa 1906, 25.

64 *Berwick Advertiser*, 9th July 1875, 1.

Following the Education Act of 1872, a campaign to introduce gardens into the elementary school curriculum began in earnest and gradually became a movement. It was around this time that school gardens started to appear in Sweden, Austria, France and Canada and articles about them, extolling their value, made frequent appearances in local newspapers. It took a long time for the idea to gain widespread acceptance in Britain, in spite of the influential voices of the radical MP Jesse Collings and the Scottish biologist, ecologist and educationalist Patrick Geddes, amongst others, and encouragement by the Board of Agriculture (set up in 1889). By 1902 there were 30,000 in France and only 100 in Britain.[65] Scotland appeared to have been even more behindhand, with teachers divided on the issue. The secretary of the Dundee branch of the Educational Institute of Scotland, the schoolteachers' union, argued that school gardens were too expensive and that 'the craze for nature knowledge would soon pass away'.[66] Only a few schools got gardens going around this time ('crops for boys, flower plots for girls ... the work ... taken up by the children with great heartiness' in Aberdeenshire.[67])

It is quite possible that some rural schoolmasters, sceptical of the school garden proposition but in need of help in their own large gardens, which were still part of their wage, were not averse to a little child exploitation. Recalling school days in Braco, Perthshire, at the end of the century, one memoir relates how, when the parish schoolmaster moved house, he was 'suddenly ... struck by the great need for the bigger boys to learn gardening. We were rather cynical about this departure from the existing curriculum but during spring and summer we were several times per week marched to his garden by the master where we were given graips and spades to clear away the weeds', which was hard work in the overgrown plot. Although the children looked forward to the more interesting sowing and planting stages, this was done on a Saturday without them.[68]

It was the 1908 Scottish Education Act that widened the curriculum, embraced vocational training and made provision for school gardens in rural areas (Plate 25). The Agricultural Colleges in Edinburgh and Aberdeen were tasked by the Board of Agriculture with the responsibility for organising them in the east and north, while the Board supplied free seeds and plants. In spite of some difficulties in acquiring land and objections by some ratepayers, school gardens now took off. Within three years the number of those supervised by the East of Scotland Agricultural College had increased from seventy-three to 128.[69]

---

65 Speech by Patrick Geddes to the Edinburgh Branch of the Educational Institute of Scotland, *The Scotsman*, 6th October 1902, 4.

66 *The Scotsman*, 26th November 1900, 7.

67 *Aberdeen Press and Journal*, 24th July 1906, 3.

68 McIldowie, *The Parish of Ardoch*, 22.

69 *The Scotsman*, 8th September 1911, 9.

Plate 25 Working in the school garden at Lilliesleaf, Roxburghshire, 1914–18.

What was their purpose – training in gardening or nature study? Patrick Geddes' idea was that school gardens would introduce city children to nature, rather than to gardening. A teacher who had already been teaching nature study in a village school, in an eloquent letter to *The Scotsman* had 'come to the conclusion that the utilitarian side [of school gardens] should not be pushed to the exclusion of the educational. Let the gardens be used for observation ... in reality an out-of-door school.' At his school, the garden, used in this way, had made pupils more interested and willing to learn and school life pleasanter.[70] However, although lip-service was paid to their potential for science education (and personal growth), those that were established by the Act were clearly mainly designed to give boys 'practical instruction' and prepare them for a future in agriculture and horticulture – 'rural pursuits'.[71]

How much training was provided for teachers is not clear. In England at this time, where between 25,000 and 30,000 boys were learning to garden at school, there was a shortage of trained teachers, and the attitude of the authorities was that anyone could teach gardening.[72] In Scotland there were some summer schools for teachers and by 1911 at least one 'experimental school garden' for training rural schoolmasters, at Mains in Dundee, with hothouses, a potting

70 Ibid., 13th August 1908, 6.

71 *Dundee Courier*, 1st July 1907.

72 John Weathers, *A Practical Guide to School, Cottage and Allotment Gardening*, London 1908, 14.

shed, a lecture hall, a rockery and a pond.[73] School gardens seem to have disappeared after the First World War, though the author remembers them at a private primary school in Edinburgh in the 1950s.

THE BOTANISTS

From the time of *The Cheap Magazine* the gardening press had offered working-class gardeners an eclectic mixture of gardening tips and snippets of botanical information. This diet, however scanty, fed the growing appetite of workers for scientific knowledge and understanding while more nutritious fare was offered by evening classes at Mechanics Institutes and Mutual Instruction associations. These last were self-help societies which began in the 1830s and spread across Britain.[74] Groups of young working men and very occasionally women met to discuss topics they had chosen and then studied, for their 'mutual instruction and improvement'.[75] Most were to be found in towns and cities, but they were very popular in the north-east countryside, where there were nineteen in 1851 and up to fifty by the 1890s.[76] Many associations had libraries and the Alford, Aberdeenshire, association library held several gardening titles and books on botany, amongst the standard biographies, travel books and sermons (and no novels).[77] The Aberdeenshire weaver–botanist John Duncan was a founding member of the Mutual Improvement association in the weaving village of Auchleven, and gave lectures on practical gardening and botany. His aim was to encourage his listeners to acquire 'a thorough knowledge of the constitution of plants, without which no correct idea can be formed of their proper purpose'.[78]

Tradesmen, skilled workers and small farmers formed the core of these associations, but the one group that appeared largely untouched by the spirit of self-improvement were the farm servants, even in Aberdeenshire. Strenuous efforts were made to recruit them with the aim of raising their moral and their physical condition, to no avail.[79] Just as they resisted the attempts of landowners to improve their gardening practice by not entering best-kept garden competitions, so they withstood other efforts at uplifting, even by those nearer

73 *Aberdeen Press and Journal*, 18th September 1911, 6.

74 Jonathan Rose traces the origins of this movement to Kinnesswood, by Loch Leven, where a group met in the home of the weaver Alexander Bruce (*Intellectual Life*, 59).

75 Topics ranged widely from current affairs to science, but practical subjects and basic literacy were taught as well.

76 Rose, *Intellectual Life*, 60.

77 Alford Mutual Instruction Association, Library catalogue, *Unity is Strength*, Aberdeen 1908.

78 Jolly, *The Life of John Duncan*, 266.

79 Carter, 'The MI movement in NE Scotland', *Aberdeen University Review*, vol. XLVI, 4, no. 156, 383–400, 387.

in social status. 'I was the only farm servant member of this Association', recalled William Milne of his time in Laurencekirk.[80]

Botany was popular throughout the Victorian era amongst all classes. Field or Naturalists Clubs were formed in Scotland from the 1870s, mainly in cities and large towns but some in villages (as in Berwickshire) had workers amongst their members.[81] Newspapers of the time are filled with advertisements for one-off lectures, series of lectures, lectures for ladies and lectures for young ladies, and lectures in villages and small towns, in the local school, at Mutual Improvement association meetings and at YMCA and YWCA meetings. In Longforgan, Perthshire, Mrs Armistead of Huntly Castle organised a series of twelve lectures on botany for the local YWCA, starting with 'insectivorous plants'.[82]

From around 1890 until the First World War, free 'Lectures on Botany for the Working Classes', for men and women, were delivered at the Royal Botanic Garden in Edinburgh by the visionary Regius Keeper, Professor Isaac Bayley Balfour. The lectures were held on Saturdays evenings, to be accessible to all. The Botanic Gardens, at that time on the edge of Edinburgh, would have been within the reach of rural workers and the lectures and their longevity are a strong indication of the eagerness for education amongst working people.[83]

The stories of two working-class botanists illustrate how love of gardening led to a fascination with the natural world and the acquisition of scientific methods. Long before the full flowering of working-class interest in botany, George Don (1764–1814) was collecting and classifying plants in his native Angus, in particular alpine plants from Glen Clova, on the edge of the Highlands. The son of a currier (a dresser and dyer of leather) who was born on a small farm near Forfar, Don was educated at the parish school and then, at a very young age, apprenticed first to a shoemaker and then a clockmaker. At fifteen he worked with his uncle at Dupplin Castle gardens in Perthshire and then for eight years as a gardener in England, spending all his spare time in collecting plants. In 1797 he acquired a plot of land on the edge of Forfar, where he built a cottage and started a botanic garden and a market garden, soon earning a reputation for the hardy plants he sold and becoming acquainted with leading horticulturalists such as Patrick Neill. His contacts took him to Edinburgh and the post of Superintendent or Principal Gardener at the Botanic Garden. While there, he published papers on the native plants of Angus in the *GV Angus* and on grasses, and, with the support of Patrick Neill, produced the first volume of the *Herbarium Britannicum*. Don became a member of the Linnean Society and attended medical classes at the University but soon returned to Forfar, having

80 Milne, *Reminiscences of an Old Boy*, 238.

81 *RC on Scientific Instruction and the Advancement of Science, Minutes of Evidence*, vol. II, 1875, 251.

82 *Dundee Courier*, 12th February 1894, 5.

83 See advertisements for these lectures in *The Scotsman*, 1890–1914.

fallen out with the Regius Keeper, and to his botanic garden and a new nursery. The nursery was never a financial success, partly owing to poor soil, partly to the high cost of labour during the war years, but also because of Don's passion for botanising. His large family was on the brink of destitution when Don died aged fifty. Two of his surviving sons became noted botanists, including George Don junior, and the other three gardeners in England.[84]

'Don possessed that divine fire ... which kept him unwearied at labour; plodding ... over the spongy morass, or breasting a high and solitary moorland; or anon climbing with all a fowler's zeal up the high rocky crags of that lovely district of Clova.' He had such a keen eye and ability to classify that he could differentiate varieties 'which unimportant as they may be to the armchair botanist, may and often do prove of material service to the agriculturist or gardener'.[85] He discovered at least three new species: the sea pearlwort *(Sagina maritima Don)*; the sea radish (*Raphanus maritimus Don*) and the alpine forget-me-not (*Myosotis alpina Don*).[86] Just as importantly, he early on identified the threat to wild plants from agricultural improvement: extraction of marl and peat caused the extinction of one species and draining wetland, he warned in 'Native Plants and Animals of Angus', risked the survival of sedges and grasses.[87]

John Duncan, a working weaver all his life who spent every spare moment studying plants and aquiring scientific knowledge, was another who may be said to have been touched with divine fire (Plate 26). Born to a single mother in Stonehaven, his childhood followed a familiar path: no schooling, a job as a herd and at the age of twelve an apprenticeship with a weaver. He then began to study herbalism with the help of Culpeper's *Complete Herbal*, having been taught to read by the local schoolmaster and friends and neighbours. For the rest of his life Duncan made a very poor living as a weaver in the Vale of Alford in Aberdeenshire, and at the same time practised as a respected herbalist, collecting plants, raising them in his flower garden and making his own decoctions, which he dispensed to his friends and neighbours. But this is only part of his remarkable story. Through his friendship with a neighbouring gardener he became interested in botany and began to study it seriously, scouring the countryside of the Vale of Alford and building up an impressive herbarium. He also taught himself Latin, Greek, Old Scots, astronomy – he was known in country places as 'the star-gazer' and made pocket sundials – as well as geology, history, biography and geography. Later in life he added theology, meteorology and entomology. His studies always had a

[84] 'George Don senior 1764–1814', available at http://www.forfarbotanists.org/georgedon.html; Alan Elliot, 'George Don sr 1764–1814', *Botanics Stories* 2013, available at http://stories.rbge.org.uk/archives/4218.

[85] G. C. Druce, 'The botanical work of George Don', *The Scottish Naturalist*, vol. II, 1885–6, 12–20, 17.

[86] Alan Elliot, 'George Don sr 1764–1814', *Botanics Stories* 2013, available at http://stories.rbge.org.uk/archives/4218.

[87] Ibid.

practical bent and he also built up a considerable store of gardening knowledge: in summer months, when weaving work tended to dry up, he often worked as a gardener. He studied gardening books and befriended many professional gardeners. On his death in 1881 his herbarium of dried specimens, containing three quarters of all British flowering plants, was gifted to Aberdeen University.[88]

His biographer, William Jolly, came to know John Duncan in his last years when he was living as a lodger on a croft at Droichsburn, near Alford. Here he had a small fenced-off piece of the garden and the use of the crofter's flower borders where he raised garlic, double wild roses and berry bushes, amongst a huge range of wild plants. When he ran out of space in his garden he sowed his specimens, collected from all over Scotland, on either side of the burn, for half a mile down to the road.[89]

Amongst John Duncan's practical advice to gardeners, he recommended forming a 'trinkie', a small circular trench, around a newly planted shrub to allow water or fertiliser to reach the young roots or spongioles. He explained how to strike new shoots by bending young branches of a tree into a mound of earth until they rooted, when they could be severed from the parent tree and planted.[90]

John Duncan cut an eccentric figure, to the point of embarrassing his friends on his visits to Aberdeen. In his fifty-year-old home-made suit, with a high collar and brass buttons, trousers rolled up his calves, heavy tacketty boots, top hat – the oldest part of his outfit – 'set sloping on the back of his head', 'his "Sarah Gamp" blue umbrella under one arm, and a large bundle under the other, and generally with a collection of plants otherwise disposed',[91] his appearance inevitably drew stares and laughter. His relations with his neighbours were often fraught as he attracted derision and suspicion for many of his activities. For all this, his great knowledge and practical skills in herbalism and gardening and his eagerness to share these, combined with his honesty, kindliness and forbearance, won him a wide circle of friends in many walks of life.

Today, the ruins of the isolated little croft beside the Droichsburn have almost vanished under a jungle of gorse, but where John Duncan's garden would have been and on either side of the burn running down to the road where he planted his specimens, a plantation of mixed native trees has been established. Duncan would have approved.

---

88 Jolly, *The Life of John Duncan*, 414–16.

89 Ibid., 321.

90 Ibid., 302.

91 Ibid., 359.

Plate 26 John Duncan, 1794–1881. William Jolly 'The Life of John Duncan, Scotch Weaver and Botanist'.

CHAPTER 9

THE IDEA OF THE COTTAGE GARDEN

What did the rural worker's cottage garden mean to those who concerned themselves most with them? Landowners, farmers, horticulturists, philanthropists and not least the workers themselves all had different although often overlapping perceptions of the function and purpose of the garden. These perceptions and beliefs weave an economic, political, social, moral and emotional web binding the different interests together. This chapter teases out the main threads of this web by examining the ideas and the motivation of the improving landowners of the eighteenth century; the horticultural and agricultural writers of the early decades of the 1800s; the philanthropists of the mid-Victorian period; and the land reformers. The voices of the workers are heard only faintly in this genteel hubbub.

Only the improving landowners and their tenant farmers materially affected the gardens of the rural workers directly but all these groups influenced, if they did not manufacture, the moral climate in which cottagers struggled to create and maintain their 'little portions of the land'.

ENLIGHTENMENT THINKERS

In his influential *Elements of Criticism* (1762) the legal theorist and philosopher Henry Home, Lord Kames (1696–1782) (Plate 27), included gardening amongst the fine arts which, he argued, it was the duty of man, especially the wealthy, to cultivate.[1] Moreover,

> no occupation attaches a man more to his duty than that of cultivating a taste in the fine arts: a just relish of what is beautiful, proper, elegant, and ornamental, in writing or painting, in architecture or gardening, is a fine preparation for the same just relish of these qualities in character and behaviour.[2]

---

[1] Henry Home, Lord Kames, *Elements of Criticism* [1762], 6th edn, Indianapolis 2005, 13–14.

[2] Ibid., 17.

Plate 27 Henry Home, Lord Kames (1696–1782), by David Martin, 1794.

Kames was a leading member of the coterie of philosophers, economists and social theorists who created the 'Scottish Enlightenment'. A largely self-taught lawyer and jurist, he published on the law and the science of human nature and was also the patron of Adam Smith, David Hume and others. In keeping with his philosophy, Lord Kames was an energetic and fiercely practical gardener with a relish for improvement. At his estate at Blair Drummond he transformed the policies from 'a bleak barren waste, without a shrub ... [to] the most luxuriant scenes of fancy' by uniting 'art to nature, profit to pleasure, and judgement to taste'.[3] It was here, too, that he began draining the huge moss by floating it down the river Forth and, by leasing small lots, rent-free at first, populated the newly emerged cultivatable soil with small farmers.

His prescriptions on gardening for landowners in *Elements of Criticism*, which he saw as bound up with the formation of a moral citizen, were in a rather different tone to his precepts on the economic value of the 'kitchen-gardens' of his tenants and labourers. *The Gentleman Farmer* (1776), written when he was eighty, was based on his own experiments and observation. In a highly critical review of all that was amiss with Scottish farming practice at the time, Kames argued:

> Thirty years ago, a kitchen-garden was an article of luxury merely, because at that time there could be no cheaper food than oatmeal. At present, the farmer maintains hisservants at double expense, as the price of oatmeal is doubled: and yet he has no notion of a kitchen-garden, more than he had thirty years ago.[4]

The farm kitchen-garden was in fact no luxury but 'the chief accommodation of all'.[5] But for Kames, the mere saving of money by the farmer was not the only benefit to be gained. As a leading and very active Commissioner for the Forfeited (Annexed) Estates, he endorsed larger views of the economic value of the kailyard. The forfeited estates were those confiscated from supporters of the Young Pretender, Prince Charles Edward Stuart, after the failure of the Jacobite Uprising of 1745. The very ambitious purpose of the Commission, which functioned from 1754 to 1784, when the estates were disannexed and returned to private owners, was to make them profitable, use the revenue to build roads and bridges and other infrastructure and improve agricultural practice. Out of the idle, ignorant and dangerous inhabitants was to be created a population of industrious and productive tenants and artisans who would happily abandon any lingering attachment to Jacobitism.[6]

The Commissioners formulated two connected strategies for this last objective and both of these crucially involved gardens: ensuring that tenant farmers, now with larger, commercially viable holdings, had access to day labourers and 'artificers'; and establishing new villages. For the former, one eighth of the land of a tenancy was to be reserved for houses, gardens and a few acres of grazing and arable for day labourers and tradesmen (weavers, wrights, smiths, masons and so on) to keep them on the land, 'to excite their industry and

3 *OSA* Kilmadock, Perthshire, vol. XX, 45.

4 Home, *The Gentleman Farmer*, 10.

5 Ibid., 274.

6 Annette M. Smith, 'The Forfeited Estates Papers – A study of the work of the commissioners for the Forfeited Annexed Estates, 1755–1784, with particular reference to their contribution to the development of communications in Scotland in the eighteenth century', unpublished PhD thesis, University of St Andrews 1974, 5.

to fix their Residence'. To be known as King's Cottagers, they were to pay their rent directly to the Commissioners. General Skene, the Inspector of Roads for the Commission, had done this on his Highland estate and his plan was submitted to the Board. Skene's plan shows a rectangular three-acre plot bounded on two sides by roads, one of them the 'King's Highway'. Two acres are allowed for grain (bere and oats) and three roods (three quarters of an acre) for grass and fodder crops. The remaining rood was to be taken up by 'a small commodious house with a cow byre and a few falls for garden stuff'. On Skene's plan the garden by itself measures approximately thirteen falls, a relatively small plot.[7]

The Board also embarked on the creation of new manufacturing and fishing villages. This was not a new idea: planned villages had been around for some time and the duke of Perth, one of the attainted proprietors, had feued land in Crieff and Callander to artisans and linen weavers well before the Forty-five. But as well as encouraging fishing and industry – that is, linen spinning and weaving – the villages, by herding the inhabitants into settlements built at road junctions, were to have a policing and supervisory role. An anonymous manuscript, *Hints Towards a Settlement of the Forfeited Estates in the Highlands of Scotland*, clearly outlines the thinking behind these villages. They would contain churches, prisons, schoolhouses and factories, as well as well-built houses. These congregations of houses

> will give the Stewards, Factors, and different Masters convenient Opportunities of instructing them in one Body in manufacturing & other Mechanical Arts; Whereby a Spirit of Industry & Emulation will be propagated amongst them; Especially as they will all be easily overlook'd and deprived of their usual Recesses for Sauntering and Slothfulness.[8]

Discipline and re-education by factors and ministers were as central to the project as higher rents.

Very similar villages, or 'colonies', were soon in demand for the hundreds of demobbed soldiers returned from America at the end of the Seven Years War.[9] Strelitz and Borland Bogg on the duke of Perth's former estate were two of these. Strelitz (named in honour of George III's queen, Charlotte of Mecklenburg-Strelitz) was built in 1763, and it was hoped it would also attract manufacturers and tradesmen. Strelitz was laid out, as the *OSA* approvingly records, 'in a commodious manner, after a regular plan, forming a spacious street, ninety feet broad, watered by a small stream, which runs along the side of the street'. Each of the eighty houses had a small garden and three acres of land enclosed by hedges and ditches and, by the 1790s, sheltered by treebelts. The inhabitants

7 NRS, E730/37/3, Commissioners for the Annexed Estates: Papers relating to Improvements.

8 Quoted in A. Smith, 'The Forfeited Estates Papers', 60.

9 Ibid., 64.

paid only a nominal rent.[10] Strelitz was inspected by Archibald Menzies for the Commissioners a few years after it was founded. He reported that the existing shelter belts were thriving, as were the hedges; the village was an 'ornament to that part of the country', was kept neat and clean and enjoyed good crops in the lower part. He recommended more tree planting, both for shelter and for fuel, both around the village and along the street.[11]

By the time of Andrew Wight's visit in the late 1770s, also on behalf of the Commissioners, however, things had begun to deteriorate. All the inhabitants were poor and by now ageing, so many were unable to work their land. Some outright refused. The land itself, according to Wight, was 'mean soil' and because of a shortage of manure and of labourers to till the land in time for sowing at the proper time, the crops were scanty. The thorn hedges were full of gaps and overgrown.[12] Lord Kames sounded the alarm for Strelitz and called on the Commissioners to address the grave situation there by reconfiguring the land as King's Cottages, giving each householder a quarter-acre kitchen garden.[13] This was agreed. The rest of the arable land of the village was to be cultivated by an 'undertaker' and the pasture was to be held in common. The alternative was beggary. Significantly, this change meant the villagers would no longer be completely independent, the original intention, as they would only be able to keep a cow and grow vegetables. The Commissioners carried out Kames' proposal,[14] but its fate is unknown. As the soldiers died or left, local people took over the empty houses and land, but by the 1820s it had been abandoned and the land planted with trees. Strelitz had from the start earned, whether deservedly or not, the scorn and suspicion of the surrounding countryside, which was apparently impressed only by the inhabitants' capacity for smuggling and drink.[15] Today only the name survives, attached to a farm and a plantation.

Strelitz failed because the soil was poor, supplies of manure inadequate and the plots of land too small for independence. The soldiers had at best limited experience of gardening or farming, and many were much too disabled to work the land. They also, crucially, lacked the financial resources to make a success of their smallholdings.

---

10 *OSA* Cargill, vol. XIII, 538.

11 NRS, E729/8, Archibald Menzies, 'Report on Perth estate 1765–7', Annexed Estates Papers.

12 Wight, *Present State of Husbandry in Scotland*, vol. I, 17.

13 In the same memorandum, Lord Kames urged that every tenant on the annexed estates be obliged to lay out a kitchen garden of at least one acre on suitable land for 'vegetables, kail, cabbage, turnip, potatoes etc.' Given the rising cost of oatmeal, 'our people would live much cheaper by dealing more in the kitchen-garden'. To encourage them, they should be given 'a good spade, shovel and hoe' ('Hints by Lord Kames to the Commissioners of the Annexed Estates', in Andrew Wight, *Present State of Husbandry*, vol. I, 187).

14 Ibid., 200.

15 Hunter, *Woods, Forests and Estates*, 354.

The £5 loan granted by the Commissioners for stocking and tools was often eaten up in food and accommodation (and drink) before they even arrived at the site.[16]

The King's Cottager scheme nevertheless helped to establish the cottage garden in Scotland as an essential instrument of improvement: without an adequate garden, these projects recognised, workers, whether in scattered cottages out in the countryside or in the often remote semi-moorland construction sites where the villages were going up, would have had no access to fresh food and no ability to maintain themselves and their families when wages were not to be had.

The ambition of the Commissioners, if not their success, which overall was very limited, undoubtedly inspired private landowners all over the country to lay out their own planned villages. Motivated more by economic than by social engineering considerations, the planned village was an improvers' solution to a problem caused by improvement. The 'Lowland clearances' had drained the countryside of many of the labourers and the artisans needed for agricultural improvement and for the creation of a new rural infrastructure.

The popularity of the new planned villages spread rapidly from the 1760s, being enthusiastically promoted by the *General Views* and many other authorities. Nearly 500 were built throughout Scotland, mostly by landowners on their own estates. Around a quarter were built on the coasts as fishing communities; some – usually very small – were designed just for weavers, especially in Angus and Perthshire. The best-known and some of the largest and most successful were in the north and north-east, where a fashionable and competitive frenzy of village-building went on in the latter years of the eighteenth century, including the grandiose but unsuccessful New Leeds and several named after their proud founders (Cuminestown, Archiestown).[17]

Douglas Lockhart has shown that the landowners' purpose in creating the villages was threefold: to attract skilled and unskilled workers for farm work of all kinds, and weavers; to establish a local market for their tenants' produce; and to increase their rent revenues.[18] A desire to minimise the disruption caused by the cottar clearances and to provide housing, employment and a patch of land for the poor may have also been in the minds of some landowners.[19] There were also aesthetic considerations: some villages, such as Tyninghame in East Lothian and Kinrossie in Perthshire, were built at least partly to beautify their owners' estates, as can still be seen today (Plate 28).

Along the straight main street, settlers feued or rented plots – 'stances' – for a house and garden. As the village grew, parallel streets were added, with cross streets connecting them. The edge of the settlement was usually waste or marginal land feued or rented to the settlers and designated for pasture or field

16 A. Smith, 'The Forfeited Estates Papers', 64.

17 Lockhart, *Scottish Planned Villages*, 11, 13.

18 Ibid., 7.

19 Devine, 'Social stability', 59.

Plate 28 Kinrossie, Perthshire.

crops such as oatmeal and barley. Rents per acre, reduced or discounted for the first few years or so, were usually soon higher than for neighbouring farmland. In New Pitsligo, founded in 1787, income from feu duty for the stances ('tenements') alone rose by a third between 1807 and 1831 and rent from the lotted lands from £6 in 1794 to £295 in 1807.[20]

Farmers benefited both from the opportunities to sell their produce in local markets, especially where the village was far from towns, and from the readily available pool of labour. As well as day labourers, tradesmen, and weavers in particular – never able to rely on full-time work throughout the year – could help out with the harvests and at other busy times (turnip hoeing in early summer, for example), or send their wives and children to the fields. There was, too, a moral or social discipline aspect to planned villagers, although lowlanders had long been civilised and never enjoyed the terrifying reputation of highlanders. Many, like the Forfeited Estates Commissioners, believed that a village community under the eye of the minister, the dominie and the kirk elders would keep the working population 'virtuous and respectful'.[21]

We have an insight into the thinking behind one of the most successful villages from Lord Gardenstone, founder in 1768 of Laurencekirk in Kincardineshire (Plate 29). Gardenstone, another member of the Forfeited Estates Commission, seems to have learned well from its failures. Settlers, mainly weavers and tradesmen, had to build their own houses and this 'effectually excluded the idle and destitute, who infest many of our villages'.[22] While not losing sight of the

20 Lockhart, *Scottish Planned Villages*, Table I.

21 Millman, *The Making of the Scottish Landscape*, 156.

22 Lockhart, *Scottish Planned Villages*, 48.

Plate 29 Francis Garden, Lord Gardenstone (1721–1793), by Henry Raeburn (circle of), no date.

economic benefits that would accrue to himself, Lord Gardenstone insisted on the independence of his settlers, although it is not altogether clear what he meant by this (not independent of his estate, anyway):

> My view from the beginning was to make the people who settled in the village independent, not doubting that such people would make my adjoining land valuable. I could not carry my land to the gates of a thriving town – but I could answer the same purpose by erecting and establishing a thriving town in the heart of my lands. – By this time I felt an agreable zeal in the project and contracted a fond affection to the people as they became inhabitants of my village.[23]

Laurencekirk was noted for its gardens from the outset. Andrew Wight, visiting in the 1770s, noted that 'many, wishing to live under so good a master, have taken feus, made kitchen-gardens enclosed with hedge and ditch, and dressed their small fields. Their houses are neat ... and everything has an air of prosperity.'[24] In 1780 Lord Gardenstone published his 'Letter to the people of Laurencekirk' on the occasion of its becoming a burgh of barony, which gave it the right to hold markets. 'I often rejoice in the report of travellers', he wrote, 'who with much approbation, observe these visible marks of your industry. The neatness of your houses, the beauty and fertility of your gardens, and the proper cultivation of your little plots of land ... are sure proofs of good sense and habitual industry in people of your condition.'[25] Gardening remained a popular activity in Laurencekirk well into the nineteenth century:

> The habits of the people are cleanly: their houses, for the most part, are kept in good order, undergoing generally more or less of a weekly purification: and it both improves the appearance of the place, and promotes the health of the inhabitants, and their taste for order, that there is attached to almost every dwelling-house a small piece of garden ground, in keeping which neatly, neighbours have a spirit of becoming emulation.[26]

Settlers were recruited by advertisement and inducements. Gardenstone offered five guineas as a premium for the first four looms to be installed in a weaver's house.[27] A 1796 advertisement appealing for settlers to New Pitsligo, the largest planned village in Aberdeenshire, included twenty-six premiums, with a total value of seventy-two guineas, and ten silver medals, to be awarded for, amongst other achievements, enclosing and cultivating moorland and installing the first five

23 Ibid., 47.

24 Wight, *Present State of Husbandry*, vol. II, 2.

25 Francis Garden, Lord Gardenstone, 'Letter to the people of Laurencekirk', *Scots Magazine*, March 1780, 148–50, 149.

26 *NSA* Laurencekirk, vol. XI, 137.

27 Lockhart, *Scottish Planned Villages*, 47.

looms. But there was more: 'the person who shall build and occupy the first slated house in the village, of two stories' would receive six guineas, the person with the most beehives in the best condition was to be awarded a silver medal and two guineas, while for 'the person who has the best built Garden Walls of Stone, the cleanest Garden, and the most thriving Fruit Trees and Bushes' there was a two-guinea prize, with one guinea for the runner-up. The proprietor, Sir William Forbes of Pitsligo, promised to make a present to every feuar of fruit trees and bushes for his garden. Forbes had drawn up his list of premiums with the assistance of the professor of agriculture at Edinburgh University, Dr Coventry. 'The medals will cost but a trifle and if those who gain them could be brought to wear them with a bitt of ribbon at their buttonhole at the kirk on Sunday, it would become an honourable mark of distinction and would excite a spirit of emulation.'[28] Whether in fact any prizes were actually awarded, or ribbons displayed, is unknown.

Not all landowners were as charmed by their settlers as Francis Garden or as generous as Forbes. The proprietor of Spitalfields in Perthshire wrote to his agent in 1773 when his village was seven years old complaining that he had

> meant that village as a receptacle of industrious tradesmen, not an infirmary for idle or unthriving beggars ... it will require both exertion activity & examples to drive laziness out of Spitlefields otherwise both their yeards [gardens] by this time had made a more decent figure & the fields behind been better plenisht with potatoes ec ec in short the whole ground around them with the assistance they have got would have a more thriving aspect. It fretts me every time I look at it.

From the outset, this feudal superior, John Mackenzie, an Edinburgh lawyer and Writer to the Signet, had been anxious about the shortage of decent and industrious tradesmen in Spitalfields, and the threat of 'drones'.[29]

Athelstaneford in East Lothian and its inhabitants, on the estate of Sir David Kinloch of Gilmerton, offered a more cheering prospect. The *OSA* commends the 'neat, commodious' well-lit two-room houses, with their large gardens on excellent soil. The feuars could also rent extra land and cultivate them with the help of the teams of horses hired out by two of their number.

> With the produce of this land the inhabitants [387 in 1792] supply themselves with meal and potatoes, and many... keep a cow. In this manner they are enabled to live comfortably, to clothe and educate their children decently and to assist in setting them out in the world. There is no village in this country where the inhabitants have improved more of late years in comfort and convenience than in the village of Athelstaneford .... In point of situation, the

[28] Ibid., 80.

[29] Ibid., 60–1.

> village is truly beautiful. It stands open and elevated, and commands the most delightful and extensive prospects.[30]

The *OSA* report does not mention the occupations of these fortunate people, but most would have been weavers or other tradesmen. The size of the gardens and the feus ensured that they were not self-sufficient or any more 'independent' than the Laurencekirk settlers.

The issue of garden size and self-sufficiency was taken up by an 1803 essay on a planned manufacturing village:

> The ground feued by each individual ought never to exceed half a rood [one eighth of an acre], including the house steading. Nor ought it to be less. If less, it will not suffice for a kitchen garden. Yet it is safer to have it somewhat less than much more. For if it exceed half an acre, this will operate against the feuar and the landholder.[31]

In this case, the feuar would not have enough time to attend both to his loom and to his farm, and so would neglect both. The landowner, on the other hand, could find his best land, if feued in larger lots, 'frittered away into shreds' and tempting the feuars to sublet their land at extortionate rates.[32] Moreover, no employer would be keen to hand out work to weavers thirled to oversized gardens. A fine balance needed to be struck.

Most of the planned villages, and the smaller feus laid out next to older villages, have survived to this day in recognisable form. Layouts followed the same principles: the main street linked the village with other towns or with the king's highway; a square in the middle or at one end was provided for the market, church and inn; side streets at right angles connected with one or more streets parallel to the main street. Houses are usually right on the street and often in terraced rows – 'without lanes or passages except the necessary gates through the house to the close or garden' to be enclosed by walls four feet high, declared the regulations (1797) for New Aberdour in Aberdeenshire.[33] Front gardens were rare – space was saved and dung hills were forced round the back – but sometimes there is a green strip between the footpath running along in front of the houses and the road, as in Charleston, Angus, New Pitsligo, Aberdeenshire, and Dirleton, East Lothian. Kinrossie, Perthshire, and Tyninghame, East Lothian, both built for estate workers, have very wide grassy strips between the road and the houses, with communal wells sunk at intervals along them.

The New Aberdour rules required each stance to 'consist of Fifteen ells [just over fifteen yards] in front by forty-eight ells back' except when local conditions

30 *OSA* Athelstaneford, East Lothian, vol. X, 167–8.

31 Robert Rennie, 'Plan of an inland village', *THASS*, 1803, 250–66, 260.

32 Ibid., 260.

33 Lockhart, *Scottish Planned Villages*, 89.

required some variation. Fifteen yards along the street was also the extent of the stances in Charlestown of Aberlour and in New Pitsligo, where they stretched back sixty-seven and fifty-five yards respectively.[34] In New Aberdour, if the house was twenty-two feet long on the street side and fourteen feet wide – standard for the time – this would leave roughly 685 square yards for the land, or nineteen falls, a generous size for a cottage garden at this time.

It was usual to offer these lands rent-free for the first seven or eight years. At Charlestown of Aberlour the first six tradesmen to apply would have an acre of moor each rent-free for seven years, on condition of liming and dunging both it and the garden at the end of the period.[35] On the other hand, Lord Gardenstone initially extracted sixpence per fall, or £4 an acre, for the stances, almost double the average for the surrounding farmland,[36] later having to halve the rent in order to attract more settlers.[37] In most places, rents rose steadily. The settlers in Charlestown of Aberlour ended up paying 21s. for their acre for the last twelve years of their thirty-one year leases.[38] Lockhart points out that, 'in spite of the relatively high rents for these lands, the opportunity to cultivate a few acres and keep a cow was also attractive to the poor since it offered them subsistence and some income from the sale of milk and crops such as hay, oats, turnips and potatoes.'[39]

The majority of these villages remained small.[40] Some failed to attract sufficient settlers or to thrive, especially those dependent on handloom weaving, or whose inhabitants were mainly the poor and the old. New Leeds, inauspiciously erected on a 'nest of thieves, poachers and smugglers', suffered from both disadvantages and never amounted to much.[41] On the other hand, many villages flourished. By the 1830s New Pitsligo had 1,262 inhabitants, a fortnightly corn market and four cattle markets each year, and 'two respectable inns and six alehouses in the village', as well as a 'pleasing and picturesque appearance'.[42] Fochabers was 'built on a neat and regular plan, with a large square in the centre, planted on each side with trees and flowering shrubs'.[43]Even tiny St Davids in Perthshire, with only fourteen feus, was 'new and thriving'.[44]

But while the proportion of the population who lived in planned villages remained very small, they deserve their place in this history because they

---

34 Ibid., 89, 91.

35 Ibid., 92.

36 *OSA* Laurencekirk, Kincardineshire, vol. V, 177.

37 Lockhart, *Scottish Planned Villages*, 48.

38 Ibid., 92.

39 Ibid., 24.

40 Enlarging a village and building more streets would inevitably involve taking in land from the existing feus, so was strongly resisted by the inhabitants (ibid., 17).

41 Donna M. Donald, *Old Strichen, New Leeds and New Pitsligo*, Catrine 2011, 50.

42 *NSA* Tyrie, Aberdeenshire, vol. XII, 725, 726.

43 *NSA* Bellie, Moray, vol. XIII, 122.

44 *NSA* Madderty, Perthshire, vol. X, 750.

established a role for the village garden it had not had before in Scotland. These villages were the scene of the cottage gardening Patrick Neill most approved of: 'The village garden is frequently the retreat of the occupier in the summer evenings after the labours of the day, when he agreeably employs himself in observing the progress of his crops, or rectifying anything amiss.'[45] The gardeners of Laurencekirk and New Pitsligo were noted for their efforts, the latter winning a rare accolade in the *NSA* for 'the superior mode in which their gardens are now kept, and the superior crops of cabbages and turnips, &c. raised on their fields'.[46] Many of these villages, including New Pitsligo, are also recorded as having horticultural or cottage garden societies, indicating that gardening was esteemed, as much as in so many Lowland villages, by the time of the full blossoming of the societies.

THE LAND QUESTION

Horticulturists and agriculturists were amongst the most active and creative contributors to debate on the function of the cottage garden, Patrick Neill and Sir John Sinclair being the prime early examples (Plates 30 and 31). Their writings are full of practical advice directed at landowners and farmers for ameliorating the lives of their workers, prescribing the optimum size and location: 'if possible, a garden is made in front of the cottage, which has an excellent effect in point of beauty, and the garden is not shaded from the sun.'[47] It should be sited near a source of water but far from stagnant pools and marshes.

Such recommendations did not spring entirely, or even mainly, from philanthropic or aesthetic impulses, although it is clear that Neill at least regarded the garden as a source of pleasure and comfort in leisure hours, as well as a necessity. If a cottage garden was to fulfil its chief economic function, that of maintaining a labour force for farmers, it had to be productive and well cultivated so as to provide an adequate subsistence for the whole family. But too much of a surplus, as in the case of the manufacturing villages, would allow workers to become truly independent. The garden needed to be a safety net, but no more. Landowners, agriculturists and employers were not interested in the kind of peasantry that had no need of cash wages. The size of the garden was therefore a critical issue.

In England both Arthur Young and Nathaniel Kent had advocated land for the poor,[48] although from different motives. Young was concerned to discourage emigration, promote the cultivation of wastes and keep poor rates down;

45 Neill, *Gardens and Orchards*, 22.

46 *NSA* Tyrie, Aberdeenshire, vol. XII, 724.

47 Sinclair, *General Report*, chapter III, app 4, 272.

48 Arthur Young, *Observations on the present state of the waste lands of Great Britain*, London 1773, 45; Nathaniel Kent, *Hints to gentlemen of landed property*, London 1775, 247.

Plate 30 Portrait of Patrick Neill (1776–1851), by John Syme, no date.

Plate 31 Sir John Sinclair, bt. of Ulbster (1754–1835), by Henry Raeburn, c. 1794–9.

Kent to jog the consciences of the landowners. Kent proposed half an acre for a labourer's garden and three acres for pasturing a cow. As Burchardt points out, Young and Kent were lone voices until the start of the French Wars, when rising poverty compounded by bad harvests prompted a more widespread call for land for the poor.[49] The allotments that were established in England at this time averaged slightly more than half an acre.[50]

Not all agriculturists agreed. Andrew Wight, an intelligent and perceptive observer and farmer, was highly critical of the demoralising effect, as he saw it, of labourers' gardens at Stobhall in Perthshire, where despite their poverty:

> not one of them will work under 10d a day ... this evil arises from a native laziness and indolence in the people; but chiefly from the whole of them being possessed of a small parcel of land; which, although insufficient to hold them constantly employed, and far less to yield a sufficiency for the maintenance of them and their families; yet such is their laziness, pride, and stupidity, that many of them refuse to work a day's labour.[51]

Home colonisation, or the 'cottage system', as a solution to emigration was discussed during and after the war years. Should the poor be helped to become completely independent of waged work by taking over waste land and even breaking up large farms? As Burchardt notes, the creation, or, according to some views, the re-creation, of such a peasantry would challenge the entire social structure.[52]

A modest alternative to this radical proposal was proposed in a prize essay which appeared in Loudon's *Gardener's Magazine* in 1830 and which was reprinted in his *Manual of Cottage Gardening* the same year. The essay was written by Archibald Gorrie. Closely involved with the villagers of Rait, encouraging their gardening, inaugurating both the village library and the cottage garden society (and collecting their rent), Gorrie was a practical gardener and a man of many interests (some quite eccentric, such as druids). He wrote for the *Gardener's Magazine* on a great variety of topics, from how to raise heartsease to forming live fences out of larch trees, and was much admired by John Claudius Loudon. For a labouring family of six Gorrie proposed a quarter of a Scots acre (forty falls), half of which would be for vegetables and half used for cropping with oats and maincrop potatoes. This was a garden, to be cultivated part-time, not a smallholding: it would occupy the labourer for only twenty-six days a year (plus six days of the labour of a boy or girl). Why a Scots labourer could make do with a little more than half the land of his English counterpart Gorrie does not explain.[53]

---

49 Burchardt, *Allotment Movement*, 12.

50 After 1830 average allotment size fell by around one-third and by the 1870s by one half (ibid., appendices 4–5).

51 Wight, *Present state of husbandry*, vol. I, 8.

52 Burchardt, *Allotment Movement*, 13.

53 D. Gorrie, 'Essay on cottage gardens', 64.

Gorrie prefaces his essay with a reply to critics of the cottage system. His experience of the villagers of Rait has taught him that treating them as 'rational beings' and supporting their gardening in a number of ways has resulted only in honesty, sobriety, cleanliness and industry. 'Colonies' of cottages, wrote Gorrie, 'will form an excellent and useful appendage to the farm'. The younger generation could work on the farm alongside the farm servants while their parents grew food for the family and a surplus for sale. 'From such a system industrious young men of sober and industrious habits may be expected to arise.' Most importantly, they would never think of emigrating abroad but would 'only remove to cultivate our own waste lands, and form new villages'. They would fill the ranks of the army and navy, or become 'virtuous and industrious mechanics'.[54]

This rather high-flown vision does not exactly match the rest of the essay. The question Loudon set for the prize was 'What quantity of garden ground does it require to supply all requisite culinary vegetables, including potatoes, to a labourer's family, and to his live stock; the former consisting of two grown persons and four children, and the latter, at an average, of one pig, three rabbits, three hens, and three ducks?' As they do not contribute to the feeding or income of the family, flowers and ornamental shrubs are not mentioned and neither are medicinal herbs or plants, such as teasels or woad used for textile processing. The essay takes the form of a calendar of tasks, from March to November, illustrated with five plans. In March, digging, dunging and planting beans, onions, leeks, succory, lettuce, early potatoes, early peas, cabbage, cauliflower, greens, savoys, broccoli and radishes (a far greater variety of vegetables than most cottage gardeners were growing at this time) and sowing oats would take up twenty-three hours. In the first week of May, 'sow a row of scarlet runners between the divisions. Dig 2 falls, and plant with late or drumhead cabbage for the pig in winter; and 1 fall to be sown with royal dwarf or green Prussian peas and turnips'. For the whole month, with hoeing and earthing up and planting more potatoes and turnips and other crops, thirty hours' work would be needed, along with five hours by a young person. May was the busiest month. By July, gardening work would take up only five hours a week in weeding, earthing up late potatoes and planting broccoli and celery.

Gorrie provides calculations of the annual value of the crops and livestock to the cottager. Vegetables and potatoes would be worth £4 11s. 3d, while eggs and two pigs would fetch £4 5s. 2½d at the market. Subtracting the labour costs, the price of the pigs and the rent, the cottager would be left with an annual net income from his garden and plot of £2 13s. 5½d. Labourers' daily wages in the Carse at the time were between 1s. 4d (in winter) and 1s. 8d (in summer). Such a garden could not provide a subsistence – wage labour was still necessary. We have no way of knowing whether any cottager's family actually attempted this back-breaking but apparently remunerative programme. One obvious problem

[54] Ibid., 63–4.

that Gorrie does not address is that the busiest period in the garden – spring and summer – was also a busy time on farms.

Gorrie's essay makes it clear that liberating the labourer's employer from excessive poor rates, by means of the garden as a safety net, entirely depended on its size; the labour of the cottager and his family was more elastic. Discussion of this issue continued throughout this period and demonstrates the concerns of writers about the economic and moral purpose of the gardens. Even more than the inhabitants of the planned villages, both labourers and farm servants had to have gardens which would enable them to supplement their wages sufficiently to keep them fit and healthy and able to reproduce but which would not be so large as to keep them from their paid employment, or exhaust them and so render them unable to help on the farms. Neill in 1813 had estimated that a much smaller garden, of twenty-four falls (a little over one eighth of an acre), would be adequate for a farm servant's family – that is, a man employed full-time, unlike Gorrie's cottager, with a pig, but that more would be useless and even harmful: it would take up too much time and require more manure than could be provided.[55] A smaller garden would force the servant to ignore rotations and cram in plants, resulting in poor crops and impoverished soil[56] – in other words, small gardens would fail to adequately maintain families.

Evidence about the actual size of cottagers' gardens is hard to come by, but Archibald Gorrie himself drew up a plan in 1809 of the part of Rait owned by his employer showing exactly this. Rait had been a thriving village on the old main road from Perth with a provost, magistrates, many tradesmen and an inn,[57] but by the early nineteenth century the inhabitants were mainly weavers and agricultural labourers and the inn had closed. According to Gorrie's plan, apart from the schoolmaster with his statutory thirty-six falls and a couple of larger plots, the other fourteen families had gardens of less than seventeen falls, and nine had only ten.[58]

A plan of Rait in 1845 shows the other half of the village, the property of Thriepland of Fingask.[59] The houses had rectangular plots of land stretching up the hill behind them, and much smaller, more irregularly shaped gardens across the street in front. The average size of the gardens and plots together was twenty-six falls, with four agricultural labourers having thirteen falls each. The median was well below twenty falls – less than half what Gorrie proposed in his essay. We do not know what the plots and gardens were used for but it is likely that the hillside areas were sown with oats, potatoes and vegetables. The gardens across the road averaged 4.3 poles, ranging in size from 1.52 to 15.6 (the

55 Neill, *Gardens and Orchards*, 21.

56 Ibid., 26.

57 David Dorward, *The Sidlaw Hills*, Balgavies 2004, 93.

58 Perth & Kinross Archives, MS115/P1, ex Bundle 23, Sketch map of Rait, 1809.

59 Perth and Kinross Archives, MS169/3/1/12/15, Plan of the village of Rait, showing property belonging to Sir P. M. Threipland.

schoolteacher's garden).[60] These would have been much too small for growing vegetables for even a small family and may have been flower and herb gardens.

In many places, however, at least in the first half of the century, it does seem that one eighth of an acre was the norm – both on the ground and in the proposals of farming and gardening writers. Alexander Thomson's report on the settling of crofters on a piece of uncultivated moorland between Stonehaven and Aberdeen, for example, recounts that each was required to lay out a garden 'of not less than one eighth of an acre' in front of their house and to enclose it with a hedge or a wall.[61] Gardens in Neill's report, farm servants' gardens, the gardens in Robert Rennie's project were mostly this size. A gardening writer, Peter Mackenzie, a regular contributor to the *Gardener's Magazine*, urged, however, a minimum size of 1,000 square yards (nearly a quarter of an acre) – 'less is a blot on the landscape'.[62]

THE MORALISTS

Alongside the economic benefits conferred by an adequately sized cottage garden, the moral regeneration of the poor figured as equally important. It is impossible to explain, and it was at the time impossible to discuss, the role of the garden in the early nineteenth century without referring to its moral function. Lurking behind this concern on the part of the well-to-do was the persistent fear of drunkenness, family deprivation and criminality and rural unrest – this last a danger that persisted from the immediate post-war years to the fading away of Chartism after 1848. This role for the cottage garden was expressed, in crude terms, in the article by Verus published in the *Gardener's Magazine* in its first year, 1826: 'a Scotchman obtains from his garden what an Englishman, in his case, commonly seeks at the ale-house'.[63] In Robert Mudie's popular *Spring*, with rather more refinement, the cottager with a pretty and productive garden (complete with roses and honeysuckle round the door) is the one who 'never wastes his time or his money, or debases his mind and learns "the broad road which leadeth to destruction" and the contamination of the alehouse'.[64] Keeping the workman at home in the bosom of his family during his non-working (they were hardly leisure) hours was a central purpose

[60] pole (imperial) = 30 sq. yards; 1 fall = 37 sq. yards; 20 falls = 740 sq. yds; 20 poles = 600 sq. yds.

[61] Thomson, 'On the settlement of crofters', 381.

[62] Mackenzie, 'Cottage gardening adapted to Scotland', 316.

[63] 'Verus', 'On field gardening', 261.

[64] Robert Mudie, *Spring, or, the causes, appearances, and effects of the seasonal renovations of nature in all climates*, London 1837, 260–1.

of the garden. And it was here, too, that he could accustom his children to a future of agricultural labour.[65]

Closely related to its moral purpose was the garden's aesthetic function. As workers' houses were slowly rebuilt or, more economically, refurbished, and a wealth of designs and models circulated amongst improving proprietors, yet another mission was foisted on the worker's cottage and its garden: to appeal to the passer-by. One way to do this was to design and build picturesque cottages. The fashion for these, in a variety of styles from classical to gothic intended to enhance a designed landscape, had first taken hold in the late eighteenth century.[66] In the middle of the next century, while most agriculturists were at pains to stress how cheaply cottages could be built or modernised, for example with factory-made cast-iron window frames, some writers were still pressing for pretty dwellings.

In 'Hints on the Improvement of Cottages', published in the *Journal of Agriculture* in 1849, David Gorrie (a horticultural journalist and Archibald Gorrie's son[67]) condemned the neglect of cottages situated at just a little distance from large and elegant farmhouses, while pointing out that labourers who feued land and built their own cottages had homes that were 'far superior in neatness and arrangement' to farm cottages.[68] At the same time he sympathised with farmers' reluctance to spend £60 or thereabouts on a cottage whose rent over a nineteen-year lease would not cover the capital cost, rather than investing in drains which would bring an immediate gain. He went on to propose – rather perversely – extremely elaborate cottage plans, built of dressed stone, with an entrance porch, a covered verandah supported by pillars and carved wooden window surrounds and gable ends. The verandah had a practical use in protecting the walls and allowing clothes and seeds to be dried, but Gorrie seemed more interested in the exotic flowers and vines that could be raised under the overhanging eaves. Mignonettes should top the garden walls, to be crowned with stone vases at the corners, although enclosures could also be made of box or spruce hedges or wire and wickerwork entwined with climbers – the whole on display to the passing traveller.[69]

Gorrie argued that cottages should not be hidden away but always be visible, sited for scenic effect and designed to harmonise with the steading and the farmhouse. He went further: planting flowers and training climbers over their walls would transform attitudes and morality. 'A love for flowers on the part of a

65 This was a point insisted on by the ploughman essayist James Alexander: while working in the family garden, 'from a very tender age', the children would be under the eye of their parents, and their disposition and attitude to work nurtured (Alexander, 'Present condition of farm-servants', 13).

66 John Woodforde, *The Truth About Cottages* [1969], London 2007, 29–34.

67 Not the ploughman David Gorrie, author of *Passages in the Life of a Ploughboy*.

68 D. Gorrie, 'Hints on the Improvement of Cottages', 4.

69 Ibid., 11, 17–18.

cottager is indicative of refined feelings',[70] while ale houses were less numerous and had fewer customers in villages where everyone had a well-kept and decorative garden than in what he termed formal villages where only some of the dwellings had gardens and where garden layouts were 'regular and tight'. The engineer R. S. Burn recommended front gardens both so that the cottager could show off his taste and because 'all such innocent and pleasing amusements are the surest antidotes to the allurements of the public house'.[71] The development of cottagers' taste had been a theme of gardening writing since Loudon. In all his writings Loudon maintained that taste should not be regarded as the prerogative of the wealthy, but that the labourer too should be encouraged to develop his taste – for his own benefit.

Fear of pauperism and the drain on the income of the well-to-do that it implied after the Poor Law Reform Act of 1845 changed the meaning of the idea of 'independence' as it applied to cottagers. David Gorrie again: as 'an observant writer on rural subjects' he had come across no instances of a cottager becoming a pauper: 'a spirit of independence ... is cherished by every one who possesses or occupies a portion of soil however small.'[72]

James Black's 'Report on the Cottage Accommodation in the District of Buchan, Aberdeenshire' continued the theme. As factor for the Ellon estate, he could justifiably claim intimate knowledge of the cottages and their inhabitants: compared to the occupiers of the old tumbledown dwellings, well-built and well-designed cottages produced more intelligent, responsible, educated and God-fearing inhabitants. 'Connected with these [superior] cottages is generally a small garden. This useful and ornamental appendage when tastefully laid out and carefully attended to, not a little enhances the good appearance of the cottage; while its presence and culture are powerfully conducive to the refinement of the taste and feelings.'[73] This is not simply a nod to the conventions: Black's chief concern in his essay is with the moral behaviour (sobriety) and education of farm workers, for which he holds the farmers mainly responsible. Peter Mackenzie, a gardening writer, expressed the same wish in spiritual terms: cottagers needed their gardens not only for food but also for pleasure. 'He would feel unhappy without his garden which will yield him comfort all the days of his life and afford him in its cultivation a salutary preparation for a higher state of enjoyment beyond the grave.'[74]

70 Ibid., 11.

71 J. S. Burn, 'Agricultural Architecture and Engineering No. 1', *Journal of Agriculture*, July 1851–March 1853, 10–20, 15.

72 D. Gorrie, 'The claims of landscape gardening' *NBA*, 3rd January 1849, 9–10, 10.

73 Black, 'Cottage Accommodation', 96.

74 Mackenzie, 'Cottage gardening adapted to Scotland', 316.

THE PHILANTHROPISTS

If the French Revolution and the post-war radical uprisings had not been unsettling enough for the upper classes, in the 1830s the rural disturbances known as the Captain Swing riots that terrorised southern England sparked fear throughout the country. Rural Scotland in fact remained peaceful at this time, but discussion of solutions for avoiding unrest amongst the peasantry (a newly adopted term – it did not appear in *THASS* before 1812), discouraging emigration and improving morals reverberated through the Scottish agricultural press. In England the revived allotment movement, led by the Labourer's Friend Society (LFS), was an immediate and explicit response to the riots. An article much-cited in Scottish agricultural journals, by a leading figure in the LFS, Captain Thomas Scobell, promoted field gardens (i.e. allotments) as a rational way of helping poor labourers 'misled by designing rioters'.[75] In Scotland cottages and cottage gardens were the battleground on which poor labourers were to be rescued from the clutches of sedition. Archibald Gorrie's essay of 1831, for example, had the ambitious aim of creating an army of patriotic young gardeners ready to colonise the Scottish wastelands rather than Canada, as well as to serve in the armed forces. It is not unreasonable to read this and articles such as Menteath's on school gardens and cottage garden societies and the HASS initiatives to get landowners to build new cottages as at least in part inspired by concerns about the moral – and political – condition of cottagers.

The HASS took the lead in efforts to raise the tone. The liberal peer the 4th earl of Rosebery, vice-president of the Society, told members in 1837 in no uncertain terms that there was no hope of improving the slovenly habits and low moral tone of the peasantry until they had better housing.[76] Premiums for cottage design were launched but enthusiasm for demolishing cottages and building new, as Rosebery had done on his Midlothian estate, did not last and by the late 1840s improvements to existing dwellings – a faster and cheaper option – were promoted instead. The best-kept garden competitions also faltered in the 1840s, as we have seen: the 1847 'Report' of the Committee on Cottages, which proposed tinkering with the prizes to encourage more competitors, also underlined the crucial role of landowners in the competitions: 'in all cases where they have neglected to show a personal interest in [the] success [of the competitions] and where they have merely offered to cottagers their benefits, without aiding and stimulating them to work them out, their apathy has been followed by the failure of the attempt.' Proprietors who took the trouble to get cottagers to compete, encouraged them and took an interest saw the competitions succeed.[77]

---

75 Thomas Scobell, 'On the benefits of field gardens for the labouring poor', reported in Miscellaneous Notices, *The Quarterly Journal of Agriculture*, vol. IV, 1832–4, 99.

76 Anon., 'Remarks on cottage premiums', *THASS*, 1839, 527–34, 530.

77 Anon., 'Cottage Premiums', 208.

At the height of the potato famine, in the midst of the Chartist agitation and following the shocking revelations of Chadwick's *Sanitary Report*, the Scottish Patriotic Society for Improving the Condition of the Labouring Classes was formed in 1846 by a group of 'benevolent proprietors and gentlemen', most of them Edinburgh doctors, academics and clergymen. The Society, taking its lead from the Labourer's Friend Society, founded sixteen years earlier, had as one of its many objects 'to promote the Field Garden system – to improve the Cottages and Gardens of the Peasantry'.[78] A resolution passed two years later enabled 'several benevolent ladies' to use the Society to distribute seeds around the country, but its initial focus was the Highlands, where the potato blight had struck most devastatingly.[79] Here the Patriotic Society devoted its efforts to improving crofters' and cottars' farming and gardening ways. Its first address harangued them about their almost universal 'neglect of the first and most obvious principles of good husbandry' – wasting manure and seeds and using horse drawn ploughs rather than the spade – and urged trenching, weeding and liquid manure. A wide range of vegetables was recommended – Jerusalem artichokes, parsnips, peas, turnips, beetroot, carrots and mangel wurzel – some of them almost unknown in Scotland and others, such as mangel wurzel, unsuited to Scottish conditions. The Society promised monthly hand-sheets to instruct the crofters in improved methods.[80]

In spite of the patronage of Queen Victoria and Prince Albert, the support of the prime minister and the duke of Sutherland as president, the Society disappears from the public record in 1850, very soon after renaming itself the Royal Patriotic Society of Scotland. Why? It seems likely that a lack of familiarity with rural conditions on the part of the Edinburgh doctors, professors and ministers must have played a part and its main purpose was never very clear. It accumulated a portfolio of unconnected and disparate projects: educating crofters and cottars in subsistence farming, designing model cottages, granting loans to crofters, establishing allotments (in Edinburgh), improving fisheries, providing information on emigration (now to be encouraged at least in the Highlands because of the potato famine) and in 1848 supporting the unemployed of Edinburgh all fell within their remit.[81]

A couple of years later a new association, studded with even more noble landowners, many also members of the HASS, was formed. Inspired by the energetic reforming minister Harry Stuart of Oathlaw, author of a widely read and influential pamphlet exposing conditions in farm servants' bothies, it had the single aim of improving Lowland labourers' and farm servants' housing. Until the 1860s the Association for Promoting the Improvement of the Dwellings and

78 *The Scotsman*, 12th December 1846, 2.

79 Ibid., 25th March 1848, 3.

80 *Caledonian Mercury*, 2nd January 1847, 2.

81 Ibid., 22nd March 1847, 2; 23rd March 1848, 3.

Domestic Condition of Agricultural Labourers in Scotland took on the task of encouraging landowners to take responsibility for bothies and farm cottages, building new and keeping them in good repair. Behind these objects lay the two venerable motivations of discouraging emigration and encouraging morality. Better and larger cottages would not only lessen the growing attraction of a freer, more comfortable and perhaps better-paid life in the towns and cities, or in North America, and discourage frequent flitting, but would also allow separate bedrooms for boys and girls and thus reduce immorality. It appears, however, that neither farmers nor landowners were attracted to the Association in significant numbers.[82]

Gardens were not a foremost concern of this Society, but its first Annual Report rehearsed Sir John Sinclair's and Patrick Neill's adage that cottage sites should be dry, healthy, well drained and close to a water supply. And 'the cottage should be so placed that the sun may shine on it all round'.[83] The size and soil of the garden were not a consideration, it seems, but looks were: 'to have the garden immediately in front, with only a footpath to the door, lends greatly to the good order and nice appearance of the dwelling.'[84] As today, a neat front garden was a badge of respectability.

What were the results of all these exhortations and projects to improve the lives of the cottagers? The 1849 HASS report on their cottage premiums in 1839 observed that, notwithstanding the progress in agriculture, industry and commerce, the 'comforts of the peasantry' in Scotland (i.e. their standard of living) had not progressed at the same rate.[85]

---

82 Its success was in doubt, as an attack in the *Dunfermline Press* makes clear. 'Unfortunately the mass of the people know little or nothing of the existence of that society. It meets periodically under the patronage, and has its meetings graced by the presence, of the nobles of our land; but it does nothing. It has indeed collected a good deal of information about how cottages may be improved, and has prepared some good models upon which new cottages may be erected. Any one disposed to save the price of a plan may, through its instrumentality, have a choice of plans gratis. But the misfortune was that under the present law of Scotland ... the law of entail forbids improvement. ... The result was that though a few cottages were built by the more favourably placed among the more benevolent of our Scottish landlords, the Cottage Improvement Association was a society that existed chiefly upon paper. It was, no doubt, all very fine – seemed, indeed, very condescending – to observe the Duke of Buccleuch and the principal nobility of Scotland met together once a year to listen to the annual report of the benevolent parson, and to say to the Scottish peasantry, "You have our good wishes, friends."' *Dunfermline Press*, 11th October 1860.

83 *1st Annual Report of the Association for Promoting the Improvement of the Dwellings and Domestic Condition of Agricultural Labourers in Scotland*, Edinburgh 1855, 21.

84 *4th Annual Report of the Association for Promoting the Improvement of the Dwellings and Domestic Condition of Agricultural Labourers in Scotland*, 23.

85 Anon., 'Remarks on cottage premiums', 527.

Fear of raising the poor rates after the Poor Law reform of 1845 was a strong disincentive to cottage building. A pamphlet published in the early 1860s by the lawyer Alexander Kennedy makes this point. Of Lord Kinnaird, Perthshire landowner, keen cottage builder and member of both HASS and the Labourer's Dwellings Association, Kennedy wrote:

> the greater part of the clay houses and crofts, in and around the villages of Inchture, and at or near Longforgan, had been knocked down since he succeeded to his father, and ... very few erections indeed had been put up in their place, fearing that not a few people might become chargeable on the poor's rates.[86]

Another barrier to cottage building was the law of entail. More than half of the land of Scotland was held under strict entail in the nineteenth century. The Act of 1685 had been designed to protect estates from being broken up by creditors and to enable them to be passed on intact to future generations. In practice, as Adam Smith and Sir John Sinclair, amongst others, railed, it discouraged investment in estate improvement. Moreover, it hindered the creation of viable smallholdings.[87]

THE CAMPAIGNERS

Questions of the right size and the multiple purposes of the cottage garden stray inevitably, as we have seen, onto the territory of independence, a 'revived' peasantry and smallholdings. John Nevay (1792–1870), a weaver poet from Forfar, who wrote many poems about the Angus countryside, used his long poem 'The Peasant' to call for land for the poor:

> Thus should the pastoral fields of Scotland be –
> Ay, and of every land: – The peasant's arms
> Blest with the fruit of love and industry; –
> The green vales portioned out in little farms –
> The happy nurseries of honest men –
> No proud monopoly in all the rural reign.
> ...
> The country is unpeopled; – few dwell there,
> But proud monopolist, and stupid clown.
> What though the land well-cultured be, and fair?
> Without a peasantry and cottages

86 Alexander Kennedy, *Papers Social and Political and Sketches of Various Social Evils in Cities, Towns and Rural Villages in Scotland*, Edinburgh n.d., 22.

87 Andy Wightman, *The Poor Had No Lawyers: who owned Scotland (and how they got it)*, Edinburgh 2010, 31–3.

The gayest summer landscape is a desert place.[88]

In nine cantos Nevay relates the idyllic life of a young peasant and his family on a small plot of land he tends himself. Nevay himself lived his whole life in the middle of the town of Forfar, probably without any kind of garden. His poem seeks to ennoble rural workers at a time when poverty and the threat of revolt was casting a troubling shadow over them. It also connects to a longing for the land which permeates British radical working-class politics throughout the nineteenth century.

Before the 1880s, however, there is little sign of land hunger in Scottish rural workers' writings. Chartism and its land programme and the calls for home colonisation from radical politicians do not seem to have brushed their consciousness and it is hard to find calls for land reform. Alexander and John Bethune, in their essay 'Practical Economy' of 1839 argued not for an independent peasantry but for a more immediate and moderate reform – giving cottagers security of tenure. Only that would encourage workers to spend time and effort on the garden:

> The poor man who has no security for the possession of his cottage beyond a single year, never exerts himself to improve its value or to beautify its appearance; because he says to himself, 'I may plant and another possess; I may labour, and another enjoy the fruits of my exertion.' But remove the same individual to a cottage of his own, where he is under no apprehension of any interference with his enjoyment while life allows him the ability to enjoy, and his conduct will be very different. He will then exert himself to the utmost to improve every thing that is capable of improvement in his possession. His little garden will be enriched by every useful and ornamental plant that he can procure, and cultivated to the highest degree of perfection which his own capacity and the nature of the subject will admit of. The ivy or the honeysuckle will be seen ascending his cottage wall; and as far as his taste and ability extend, no care or labour will be spared, which can increase its comfort or beautify its appearance.[89]

This had also been Patrick Neill's position and one that was shared by some enlightened landowners.

The Clodpole argued that the lack of smallholdings and the low cash wages meant that ploughmen, unlike 'industrious workmen', had no chance to better themselves by their own efforts even though crofts generated higher rental incomes per acre than large farms.[90] He held up to his readers the twin bogeymen

---

88 Nevay, *The Peasant*, 24.

89 Alexander and John Bethune, *Practical Economy: explained and enforced in a series of lectures*, Edinburgh 1839, 127–8.

90 *The Clodpole*, 9.

of emigration and the sapping of military might as inevitable results of the failure of a peasant economy. The Aberdeenshire ploughman James Alexander rather more cautiously proposed '(where it can be had) a small croft' as even better than a garden for promoting industry. He had a poor opinion of his fellow servants, however, holding that ignorance and lack of instruction from masters and others often hindered them from making the best use of their plots and thus feeling 'they can do something to help themselves'. He argued that gardens too inculcated 'industrious habits' in farm servants.[91]

In the 1870s the question of allotments and smallholdings for agricultural labourers was forced onto the national agenda by the joint efforts of two charismatic figures, both English Liberal MPs: Joseph Arch, founder of the English National Agricultural Labourers Union (NALU), and Jesse Collings, a tireless campaigner for rural workers and ally of Joseph Chamberlain. From its beginnings in 1873 the NALU demanded land for allotments and smallholdings. The 1885 election, the first at which working men could vote, was dominated by Chamberlain's 'three acres and a cow' campaign. Throughout the 1880s Collings fought for legislation for allotments and for a 'peasant proprietary' to be created through state loans to labourers. He argued that the creation of an English peasantry would help to repopulate the countryside, reduce poverty and emigration and lessen the country's dependence on imported food.[92] His efforts resulted in the Allotment Extension Act of 1882, which brought about an enormous increase in the number of allotments (council-owned) in England. The 1908 Small Holdings and Allotments Act strengthened the earlier Act's provisions, allowing the compulsory purchase of land. The 1919 Land Settlement Act finally made smallholdings affordable, although the impact of even this legislation on the lives of agricultural workers was small.[93]

Collings was not alone, although by a long way was the most committed campaigner for allotments and smallholdings. As Paul Readman has shown, the image of the sturdily independent and industrious 'peasant' had an almost magical hold over both Liberals and Tories in the 1880s and up to the First World War, as the threats posed by the land agitation in the Highlands, labour unrest throughout the country and the socialist movement loomed ever larger.[94]

Collings' ideas, at least so far as the rural farm workers were concerned, found little echo in Scotland. Only the Scottish Ploughmen's Union, founded in 1887 in Perthshire, took up the call. This union's 'primary object' was to campaign for smallholdings for its members.[95] In village meetings all over Angus and

---

91 Alexander, 'Present condition of farm-servants in Scotland', 13.

92 Jesse Collings and Sir John Little Green, *Life of the Right Honourable Jesse Collings*, London 1920, 170.

93 Joseph F. Duncan, *Agriculture and the Community*, Stirling 1921, 48.

94 Paul Readman, *Land and Nation in England – patriotism, national identity and the politics of land, 1880–1914*, Woodbridge 2008, 43 ff.

95 *Dundee Courier*, 17th November 1887, 3.

Perthshire its leader Robert Duncan, a well-known Liberal, argued tirelessly for smallholdings as well as for shorter hours, holidays and more regular wages. Duncan soon had to retire due to ill health and when his union merged with the Aberdeenshire-based Scottish Farm Servants, Carters and General Labourers Unions the smallholdings demand seems to have been dropped.[96]

Allotments, too, had very little appeal in the Scottish countryside. A Scottish version of the Allotments Act was passed in 1892, but almost no rural allotments were established. According to Caitlin Desilvey, the Act was seen 'more as an irritant than as an opportunity'. A circular proposed by the secretary of state for Scotland to explain the Act to potential allomenteers was abandoned when reports arrived 'testifying that the Scotch labourer does not want an allotment'. Councils were reluctant to allocate resources and were also blamed for dragging their feet.[97]

The 1893 Royal Commission on Labour investigated allotments and demand for allotments one year after the passing of the Act. Throughout the rural Lowlands interest on the part of labourers, for whom the Act was designed, was at zero. The Assistant Commissioner R. Hunter Pringle believed, probably rightly, that

> the sole attraction in a small plot of land to a labouring man, is that thereby he is enabled to grow a supply of potatoes and vegetables and keep a cow for his cottage and family. Now, in Scotland he has these advantages without the risk or toil connected with an allotment.

Pringle questioned hundreds of labourers at public meetings in the Lothians and found 'not one voice' in favour: they were more suited to townsfolk or shopkeepers and railway servants, he concluded. In central counties smallholdings were equally unpopular – 'rather a curse than a blessing'.[98] It is not hard to work out why: while employed full-time in the farm ploughmen would not have the time to keep an allotment and once retired had no savings to provide working capital for stocking smallholdings. Some also felt the state of agriculture at the time would make such a venture too risky.

In the north there had been no demand for allotments either, but the case for smallholdings and crofts here was more compelling, and they had always been popular. Most were rented or feued by tradesmen and shopkeepers looking to supplement their incomes. Nevertheless,

> all seem to agree [on the Moray Firth coast] that, while he is in service, a farm servant cannot look after a croft, or anything more than the little patch of

---

96 Personal communication from Ian Gasse, 23rd October 2020.

97 Caitlin Desilvey, '"When plotters meet": Edinburgh's Allotment Movement 1921–2001', unpublished MSc thesis, University of Edinburgh 2001, 18.

98 *RC Labour*, vol. III, pt II, Kinross, Clackmannanshire and Fife, 61.

> garden ground, perhaps one eighth of an acre which he has. But the ambition of a farm servant is to have a croft or a small farm

– and the means to stock it.[99] Uniquely, in Blairgowrie in Perthshire soft fruit production on smallholdings rented by former labourers and others was booming: 'one of the great beauties of fruit-culture was that it enabled many who would otherwise be working for weekly wages of 20s. or 25s. a week to earn a good living by their own skill and industry', announced Captain Proctor of the Volunteers at the Annual Dinner Dance of the Blairgowrie and Rattray Fruitgrowers Association.[100]

In the years leading up to the First World War numbers of smallholdings (holdings under fifty acres) were static or in decline throughout Scotland, with the exception of the Highlands.[101] This trend was most marked in Aberdeenshire and the Moray Firth counties.[102] Nevertheless, an article in the *Scottish Farm Servant* in 1913 claimed that servants were still ambitious for smallholdings, for themselves or their children.[103] The Labour Party's Rural Programme at this time was calling for easier access to smallholdings and cheap loans.[104] Joseph Forbes Duncan, founder and secretary of the Scottish Farm Servants Union, accounted for the falling-off in smallholdings in a forceful post-war pamphlet on the future of agriculture. Two thirds of those who applied for smallholdings were not farm workers but tradesmen, miners and sailors who had been able to accumulate some capital, he maintained. Duncan poured scorn on the two notions, beloved of politicians (like Collings?), that smallholdings were better cultivated and more profitable than large-scale enterprises and that they would repopulate and reinvigorate the countryside. Only the development of educational and cultural opportunities in rural communities would truly revive them:

> all small-holding communities are overworked communities. They can exist only by excessive labour of men, women and children. Thrift becomes a vice because the petty economies must be hourly observed .... Under competitive conditions ... the smallholder must always be too self-centred to become a good citizen or the stuff from which a healthy rural community can be created.[105]

99 *RC Labour*, vol. III, pt I, 112.

100 *Blairgowrie Advertiser*, 18th January 1902, 3.

101 There were very small increases in parts of the Lothians (possibly market gardens) and the Borders.

102 Board of Agriculture for Scotland, *Agricultural Statistics 1913 vol. II, pt I*, Tables 13 and 14, 86–7.

103 C. A. Paterson, 'The housing of farm servants', *The Scottish Farm Servant*, May 1913, 16.

104 'The Labour Party's rural programme', *The Scottish Farm Servant*, July 1913, 6.

105 Duncan, *Agriculture and the Community*, 54.

Crofters in the north-east certainly found their new lives less secure, demanding of even harder toil and more poverty-struck than before. 'The stark truth of it was this: there was no independence, only a greater tyranny than the grieve ever imposed.' To make ends meet they needed to work for the big farmers, or go fishing, or take a part-time job as a postman: 'whenever his croft needed him, whenever his corn was ripe for the scythe, he was contracted elsewhere … . He lost his own harvest while working other men's fields. In time he lost hope as well.'[106]

It is easy to see why farmers often favoured crofts. The earl of Dalhousie's factor argued 'they are absolutely necessary to the estate, not as being directly a source of profit, but as making large farms more marketable.'[107] If their holdings were in the right place (i.e. near farms), crofters could be hired as casual labourers on farms and on estates or in quarries, but the crofts needed to be on good soil and not, as was often the case, on barren hillsides if they were to thrive and continue to be of use. Other farmers told the Commission that they were 'a useful reservoir and nursery of labour' and an incentive for thrift.[108] The Assistant Commissioner G. R. Gillespie bemoaned the feckless ploughmen who spent their cash on luxuries like coffee and jam and bicycles when they could have been saving for a croft.[109] Such arguments were exactly the same as those landowners and agriculturists had advanced sixty and more years earlier in favour of cottage gardens.

For much of the nineteenth century a cottage garden could be used by masters and landowners as a means of social control, benevolently or otherwise. A garden or a plot of land on good soil and in a favourable location could induce a farm servant, a miner, a labourer or a tradesman to switch employers or to build or occupy a cottage. Once settled, the family would be coaxed to make full use of the garden with slips and seeds, stimulated by the prospect of a flower show or a best-kept garden competition, or threatened with possible eviction in the event of neglect, as in the case of many mining villages. The garden was also to be a schoolroom for a future generation of skilled, hard-working and morally upright farm workers.

Jeremy Burchardt argues, in relation to allotments in England, that in the second half of the nineteenth century landowners became less interested in their transformative moral power, partly because allotments had become so common.[110] In Scotland the improvement in rural workers' standard of living, the weakening hold of landowners and farmers over their workforces and the increasing opportunities to escape altogether from the countryside all meant

106 Cameron, *The Ballad and the Plough*, 158–9.

107 *RC Labour*, vol. III, pt I, Arable districts of Forfarshire and Perthshire, 154.

108 Ibid., Arable districts of Aberdeenshire and Kincardineshire, 132.

109 Ibid., Arable districts of Forfarshire and Perthshire, 136.

110 Burchardt, *Allotment Movement*, 227.

that gardening could no longer be seen as a test of the survival of the physically and morally fittest. The garden meant less. The shrinking of farm servants' gardens, the relative failure of the Allotments Act and the agricultural labourers' rejection of smallholdings all point to a slackening of the tie to the land. But, for many cottage gardeners, passionately involved in their garden as a hobby and a creative space, John Duncan's words would have struck a sympathetic note:

> There is hardly a spot of earth so rugged in which the art of the gardener will not be found to produce something like loveliness in the scene; scarcely a tribe of man so rude among whom it will not create some idea of beauty, to lift up his mind to the Supreme Fountain of light and beauty and the Giver of all goodness; and there is scarcely a cottage so small that may not have the rose and the woodbine winding round its porch, and Clematis or virgin bower.[111]

[111] Jolly, *Life of John Duncan*, 266.

EPILOGUE

Rows of farm cottages, neat village houses of dressed stone and older dwellings built of rubble, clay-and-bool and even mud can still be found in their thousands in the Eastern Lowlands. In a myriad ways, the landscape of today is still recognisably the landscape of improvement. But no East Lothian hind or Angus weaver would recognise the gardens of their cottages today. The characteristic layout of vegetable beds separated by cinder paths, with climbers covering the walls and roofs and narrow flower beds squeezed up against the house walls, have long gone, along with the middens and the pigsties.

The Scottish countryside has become increasingly the preserve of the middle class, and for them the cottage rows have been remodelled into single dwellings with conservatories and extensions. New country dwellers often have little time for intensive gardening and prefer the idea of an outdoor living space to that of an outdoor larder. The gardens now contain driveways, garages, patios and spaces for barbecues, and lawns planted with trees and shrubs have replaced the vegetable plots. Sometimes they have been landscaped in imitation of the grounds of great houses. Other cottage gardens have shrunk as slices have been taken off for development or to enlarge the surrounding fields. On the other hand, there are some cottage gardens which have kept, or rediscovered, their role as providers of fruit and vegetables, with added polytunnels, hen-coops and beehives. Nevertheless, to travel through today's Lowland countryside is to pass through a far more prosperous, more manicured, but also less productive landscape than that investigated by journalists and government commissioners before the First World War.

The legacy of the gardening cottagers of this book needs to be sought not in today's country cottages but amongst the allotment-holders and the community gardeners, mostly urban, of the twentieth century and today. Like the old cottage gardens, allotments are designed to feed families. Like the old cottage gardens, they had originally a moral dimension – allocated to the deserving poor or, between the wars, to the unemployed. Nowadays allotments may have lost this dimension, but have been consciously transformed into havens of quiet contemplation, social interaction, mutual support and environmental regeneration.[1] Like the old cottage gardens, too, they have often come with built-in insecurity of tenure, as the land they occupy is eyed up by cash-strapped councils and profit-hungry developers.

In one important respect, allotmenteers today may be different from the cottagers. Allotments are now sociable places, where the plot-holders share advice, cuttings and cups of tea, although this may be only a recent development brought about by the changing demographic make-up of allotmenteers – most are now women. On my father's allotment in Edinburgh in the 1950s grunts were the usual greeting between the overwhelmingly male plotters, and nothing much else seemed to be shared.

All the working-class accounts mentioned in this book make it clear that, for them, gardening was a solitary or at most a family activity. In this book there have been only a few glimpses of gardeners working side by side. These apart, the extent to which cottage gardeners, even farm servants living cheek-by-jowl in a row of farm cottages, were accustomed to helping each other or co-operating over the fence is simply not known.

Gardening has, however, always had the reputation of being an individualistic occupation. The goal of independence or self-sufficiency that many of the gardeners in this book aspired to or at least were credited with by their betters seems to run counter to co-operation and to collective endeavour. On the other hand, there may have been more of this than I have discovered. It is hard to imagine that the farm servants who worked together on the farm did not also help each other out in the garden when needed or that the old women who won prizes in the best-kept garden competitions, for example, did not have quite a lot of neighbourly assistance. Shows and competitions – another enduring legacy of the past where they still survive – were above all community events, notwithstanding the essential 'spirit of emulation' they were designed to evoke (while not forgetting the lengths to which competitors such as the Killearn villagers were prepared to go). Pretty villages such as Forgandenny, Dirleton and Lhanbryde can only have been the result of both emulation and co-operation.

I would argue that another inheritance of the kailyards and cottage gardens is culinary. Scotch broth and other traditional soups made with vegetables are still enjoyed today, both home-made and tinned. Fifty or sixty years ago in Scotland they were even more popular. A few years ago I carried out a very informal survey of sixty-four friends and acquaintances in the UK, mostly brought up in the 1940s, 1950s and 1960s, in order to discover their childhood memories of home-made soup. Londoners ate only tinned soup, even in families where a good cook provided the meals. From the English Midlands north, it seemed that the further up the country someone was brought up, the more likely they were to have been fed home-made soup. This trend cut across class boundaries: all the Scots

[1] See Jenny Mollison, Judy Wilkinson and Rona Wilkinson, *Raising Spirits: allotments, well-being and community*, Edinburgh 2014.

Plate 32 Twenty-first-century farm cottages near Dundee.

respondents had home-made soup, regardless of social class. Many recalled the traditional soups their mothers made:

> My mother made soup regularly, at least twice a week – large pots which lasted for a few days. This was usually what we ate at lunchtime, which we referred to as dinner time (the evening meal, which was usually the meat/veg, fish etc was referred to as tea time).There were two main types. One was lentil soup, made with a ham end, grated carrots, onions, lentils etc. The other was called 'Kale', which was actually a thick Scotch broth. A piece of flank mutton was cooked in it. The meat from both of the soups would be served separately at tea time on the day it was made, with potatoes and other veg.[2]

The history of cottage gardens is largely undiscovered territory. Much remains to be unearthed, especially at a very local, parish or village, level. To amplify and correct the picture that I have only sketched here, more local research is needed. I suspect there is much more information about cottage garden shows in local libraries and archives, family papers and family memories than I have been able to find. I would encourage local history enthusiasts to delve into the gardening history of their villages and rural districts.

There are many reconstructed historic gardens and well-researched displays of ancient herbs in Scotland, but there is no reconstruction of a worker's cottage garden in the Scottish Lowlands. This, perhaps more than any book, would open a gate into the world of the kailyard and the cottager's garden.

2 Personal communication from Pat Fraser, 11th November 2013.

GLOSSARY

appleringy: (herb) southernwood

bere, bear: a primitive variety of barley

boll: 140lb of dry measure

bondager: a female field worker whom farm servants and labourers were bound to provide for a set number of days each year, in payment of the rent of their cottage and garden. A common practice in Lothians and Berwickshire.

bothy: a building on the farm housing single farm servants.

bothymen: single farm servants accommodated in the bothy.

burn: stream.

carse: low-lying land near a river. Originally marshy, but when drained (seventeenth–eighteenth centuries) extremely fertile. The Carse of Gowrie lies between Dundee and Perth on the River Tay, and there are carses around Stirling and Falkirk in the Forth valley.

cadger: pedlar.

chaumer: accommodation in the farm for single farm servants.

chiel(d): a friendly term for a man or young man.

clay-and-bool: mud and round stones used to build cottage walls.

coquiny: term used by John Duncan and found nowhere else. Seems to mean a small amount.

cottar, cotter: originally (pre-Improvement) a kind of subtenant on a fermtoun holding land in return for services to the tenant; later, a cottager or (in Aberdeenshire) a farm servant.

cottartoun, cottertown (also **cottoun**): a settlement of cottars (pre-Improvement).

crowdie: soft cheese made from curds.

dominie: parish schoolmaster.

dreel: drill (e.g. for potatoes).

eggler: hawker who collects eggs from farms and villages for sale at neighbouring markets.

factor: an agent who manages an estate, or property.

fall: a measure of area, thirty-six yards (linear or square).

farmer's ha': farmer's hall or dwelling.

fee: servant's wage (for a year or half-year hire); feeing fair, hiring fair.
fermtoun: (pre-Improvement) a farming settlement of tenants, subtenants and cottars.
feu: the permanent tenure of a piece of land in return for the annual payment of a fixed sum of money to the owner of the land.
feuar: one who holds a feu.
flitting: moving house.
gains: payment in kind (the term was mainly in use in the south-east).
girnal, girnel: large chest for storing meal.
graip: garden fork.
grieve: overseer, bailiff (on a farm).
gudeman: (pre-Improvement) farmer.
halfling: a teenage boy employed on a farm.
heritor: landowner in the role of contributor to the upkeep of the parish church, school and sometimes paupers.
hind: (south-east) farm servant hired for a year or six months (long hire).
hingin' lum: chimney attached to the inside of the gable end of a cottage, usually made of wood.
howk: dig up.
ilka: each, every.
infield: (pre-Improvement) the more fertile land near the fermtoun, manured and cultivated every year.
kebbuck: home-made cheese.
kailpot: cooking pot.
kailyard: vegetable or kitchen garden.
kirk session: elected committee (of 'elders') responsible for governing the parish.
kist: chest for storage.
lademan: miller's carter.
laird: landowner.
long hire: twelve- or six-month contract between the farmer and the farm servant.
lum: chimney.
orraman: general farm servant.
outfield: (pre- Improvement) less fertile land cultivated less regularly than infield and further from the fermtoun.
paffler: smallholder (pendicler).
policy: grounds or park surrounding a mansion.
potherbs: leafy vegetables for the pot, such as cabbage, kale.
rig: (pre-Improvement) a strip of ploughed land raised in the middle and sloping gradually to a furrow on either side. Later, a furrow.
runrig: (pre-Improvement) a way of apportioning rigs amongst the tenants and subtenants of a fermtoun.
shaw: stalks and leaves of potatoes and other root crops.

siccan: such a.
sowens: fermented oat husks and meal, cooked like porridge.
steading: farm buildings.
suppering: feeding (horse).
swee, swey, sway: metal arm attached to the side of a fireplace on which a cauldron or kettle can be hung.
syboes: spring onions.
tack: lease.
tacksman: one who holds a tack.
thirled: bound.
writer to the signet (WS): member of an ancient society of solicitors.

BIBLIOGRAPHY

ARCHIVAL COLLECTIONS

Aberdeen City and Aberdeenshire Archives

AS/Apcul Culsalmond Parochial Board/Parish Council Minute Books.

AS/Aplon Lonmay Parochial Board/Parish Council Minute Books.

Dundee City Library Lamb Collection

Mains and Strathmartine Horticultural and Poultry Society Prize Schedule for Competition, 20th August 1887.

Kelso Horticultural Society Records (private archive collection)

Kelso Horticultural Show Schedule 1850.

Killearn Cottagers' Horticultural Society Records (private archive collection)

Minutes, Killearn Cottagers' Horticultural Society, 1870–83.

National Library of Scotland

Acc 10706 Stevenson family business papers, Annual Reports on Northern Lighthouses.

National Records of Scotland

E729/8 Archibald Menzies, Report on the Perth Estate, Annexed Estates Papers.

E730/37 Commissioners for the Annexed Estates: Papers relating to Improvements (General Improvements) 1761.

GD18/1735/2 Clerk of Penicuik estate papers, day labour books, 1817–19.

GD172/764/2 Henderson of Fordell papers, Memorandum on arrangements to take effect at Fordell at Martinmas 1879.

GD172/2847/3 Schedule, Mossgreen and District Floral and Horticultural Society Annual Exhibition, 11th September 1880.

HH62 County Medical Officer of Health Reports 1891.

NLC 5/5/5 Northern Lighthouse Board, Instructions to light-keepers, 3rd edn, Edinburgh 1874.

Perth & Kinross Archives

MS115/P1, ex Bundle 23 Stuart of Annat papers, sketch map of Rait, 1809 (by Archibald Gorrie).

MS169/3/1/12/15 Thriepland of Fingask papers, Plan of the village of Rait, showing property belonging to Sir P. M. Threipland, 1845.

RHS Lindley Collections

RHS/VS/hist/10.2 The Handwriting of Under Gardeners and Labourers.

CENSUS ENUMERATION SCHEDULES, 1841–1901

(available on Scotlandspeople website, http://www.scotlandspeople.gov.uk and on microfilm in many local libraries)

CONTEMPORARY BOOKS AND ARTICLES

A Working Man, *The Clodpole: or the normal condition of agricultural labourers*, Dundee 1858.

'Abstract of reports on collecting and applying liquid manure', *THASS*, 1849, 253–275.

Anon., 'Premiums offered by the Society in 1829, VII Cottages', *THASS*, 1829, Appendix 1, 38.

Anon., 'Provincial horticultural societies', *GM*, 1836, 643–75.

Anon., 'Provincial horticultural shows', *GM*, 1837, 581.

Anon., 'Remarks on cottage premiums', *THASS*, 1839, 527–34.

Anon., 'Cottage premiums', *THASS*, 1849, 203–10.

Anon., 'Our agricultural labourers', *Journal of Agriculture*, July 1849–March 1851, 440–8.

Anon., 'Horticulture for cottagers', *The Gardener*, 1st August 1870, 337–41.

Anon., 'Liquid Manures', *Journal of Horticulture and Cottage Gardener*, 1871, 218.

Adam, William, Letter to the President of the Board of Agriculture on improvements on Blair, 1797, in *GV Kinross*, Appendix.

Adamson, Robert, *The Cottage Garden*, 2nd edn, Edinburgh 1856.

Alexander, James, 'Prize essay on the present condition of farm-servants in Scotland with means to be adopted for its amelioration', *North of Scotland Gazette*, Aberdeen 1856.

Alford Mutual Instruction Association, Library catalogue, Aberdeen 1908.

Allardyce, A. (ed.), *Scotland and Scotsmen in the 18th century from the MSS of John Ramsay of Ochtertyre*, Edinburgh 1888.

Lord Belhaven, *The Countrey-man's Rudiments, or, An advice to the farmers in East-Lothian, how to labour and improve their ground*, Edinburgh 1699.

Bell, James, 'Account of the Culture and Produce of a Field of Potatoes in the Vicinity of Leith', *THASS*, 1803, 471–8.

Bethune, John and Alexander, *Practical Economy: explained and enforced in a series of lectures*, Edinburgh 1839.

Bethune, Alexander, *Poems by the late John Bethune, with a sketch of the author's life*, 2nd edn, London 1841.

—, *Memoirs of Alexander Bethune, embracing selections from his correspondence and literary remains*, ed. William McCombie, Aberdeen 1845.

— and John Bethune, *Tales of the Scottish Peasantry*, London 1884.

Black, James, 'Report on the Cottage Accommodation in the District of Buchan, Aberdeenshire', *THASS*, 1853, 92–9.

Boswell, James, *The Journal of a Tour to the Hebrides with Samuel Johnson* [1785], London n.d.

Bremner, David, *The Industries of Scotland – Their Rise, Progress and Present Condition* [1869], Exeter 1969.

Buchan-Hepburn, George, *General View of the Agriculture of the County of East Lothian*, Edinburgh 1794.

Burn, J. S., 'Agricultural Architecture and Engineering No. 1', *Journal of Agriculture*, July 1851–March 1853, 10–20.

Chadwick, Edwin, *Report into the Sanitary Condition of the Labouring Population of Great Britain* [1842], ed. M. W. Flinn, Edinburgh 1965.

Cobbett, William, *Rural Rides* [1830], ed. George Woodcock, Harmondsworth 1967.

—, *Tour in Scotland and in the Four Northern Counties of England in the Autumn of the year 1832*, London 1832–3.

Collings, Jesse and Sir John Little Green, *Life of the Right Honourable Jesse Collings*, London 1920.

Curtis, C. H., *The Book of the Flower Show*, London 1910.

Dempster, Robert, *The Natural Horseman, the horse and ploughman*, Dundee 1884.

Diack, William, 'The Scottish farm labourer', *The Independent Review*, vol. VII, 1905, 315–27.

'Domestic notices – Scotland', *GM*, 1828, 355.

'Domestic notices', *GM*, 1832, 470–6.

'Domestic notices', *GM*, 1839, vol. XV, 195–8.

Donaldson, James, *General View of the Agriculture of the County of Elgin or Moray*, Edinburgh 1794.

—, *General View of the Agriculture of the Carse of Gowrie*, Edinburgh 1794.

—, *General View of the Agriculture of the County of Nairn*, Edinburgh 1794.

Douglas, R., *General View of the Agriculture of the County of Selkirk*, Edinburgh 1813.

—, *General View of the Agriculture of the county of Roxburgh*, Edinburgh 1813.

Druce, G. C., 'The botanical work of George Don', *The Scottish Naturalist*, vol. II, 1885–6, 12–20.

Dr Duncan and Patrick Neill, 'Report to the Highland Society respecting Early Potatoes', in Sir John Sinclair, *General Report of the Agricultural State and Political Circumstances of Scotland*, Edinburgh 1814, chapter IX, appendix vol. II (A).

Duncan, Joseph F., *Agriculture and the Community*, Stirling 1921.

Duthie, Jane Anne, *Rhymes and Reminiscences*, Brechin 1912.

Erskine, J. F., *General View of the Agriculture of the County of Clackmannan*, Edinburgh 1795.

Farrall, Thomas, 'On the agriculture of the counties of Edinburgh and Linlithgow', *THASS*, 1877, 1–66.

Fraser, Amy Stewart, *The Hills of Home*, London 1973.

Garden, Francis, Lord Gardenstone, 'Letter to the people of Laurencekirk', part II, *Scots Magazine*, vol. 42, March 1780, 148–50.

Gilly, W. S., *The Peasantry of the Border: an appeal on their behalf*, Berwick on Tweed 1842.

Gorrie, Archibald, 'Observations on the cottage gardens in the Carse of Gowrie', in Patrick Neill, *On Gardens and Orchards*, Appendix D, Edinburgh 1813.

—, 'Rait village library', *GM* 1827, 374–5

—, 'An essay on cottage gardens', in J. C. Loudon, *A Manual of Cottage Gardens, Husbandry and Architecture*, London 1830, 62–72.

Gorrie, David, 'Hints on the improvement of cottages', *Journal of Agriculture*, 1848, 1–27.

—, 'The claims of landscape gardening', *NBA*, 3rd January 1849, 9–10.

—, 'Cottages in park scenery', *NBA*, 24th October 1849, 10.

—, *Passages in the Life of a Ploughboy*, Dundee 1857.

Grant, Joseph, 'The warlock of Glendye', in *Tales of the Glens, with ballads and songs*, Stonehaven 1835.

Headrick, James, *General View of the Agriculture of the County of Angus, or Forfarshire*, Edinburgh 1813.

Home, Henry, Lord Kames, *The Gentleman Farmer: being an Attempt to Improve Agriculture by Subjecting it to the Test of Rational Principles*, Edinburgh 1776.

—, *Elements of Criticism: the sixth edition*, ed. Peter Jones, Indianapolis 2005.

Hope, Frances Jane, *Notes and Thoughts on Garden and Woodland Written Chiefly for Amateurs*, London 1881.

Hunter, Thomas, *Woods, Forests and Estates of Perthshire*, Perth 1883.

Hunter, T. F., *Mining in West Fife: the Crossgates area in particular Fordell and Halbeath collieries*, Norwich 2001.

Hutchison, Robert, 'Report on the dietaries of the Scotch agricultural labourer', *THASS*, 1869, 1–29.

Jolly, William, *The Life of John Duncan: Scotch weaver and botanist*, London 1883.

Kennedy, Alexander, *Papers Social and Political and Sketches of Various Social Evils in Cities, Towns and Rural Villages in Scotland*, Edinburgh n.d.

Kent, Nathaniel, *Hints to gentlemen of landed property*, London 1775.
Kerr, Robert, *General View of the Agriculture of the County of Berwick*, Edinburgh 1809.
Laing, Alexander, *Wayside Flowers*, 4th edn, Edinburgh 1878.
Lawson, George, *Reminiscences of Dollar Academy, Church and Sunday School*, 3rd edn, Alloa 1906.
Lightfoot, John, *Flora Scotica*, London 1777.
Littlejohn, Dr Henry D., *Report on the sanitary condition of the City of Edinburgh*, Edinburgh 1865.
Lowe, Alexander, *General View of the Agriculture of the County of Berwick*, Edinburgh 1794.
Macdonald, James, 'On the agriculture of the county of Fife', *THASS*, 1876, 1–60.
Mackenzie, Peter, 'Cottage gardening adapted to Scotland', *GM*, 1842, 315–20.
Maclaren, Ian, *Beside the Bonnie Brier Bush*, Edinburgh 1895.
McIldowie, Daniel J., *The Parish of Ardoch and the Village of Braco*, Braco 1981.
McNab, Robert, *The North British Cultivator: a treatise on gardening, agriculture and botany*, Perth 1842.
'Meg Dods' [Isobel Christian Johnston], *The Cook and Housewife's Manual: a practical system of modern domestic cookery and family management*, Edinburgh 1829.
Menteath, James Stuart, 'On improving cottage gardening in Scotland', *Quarterly Journal of Agriculture*, vol. III, 1831–2, 783–5.
—, 'On cottagers' garden societies', *Quarterly Journal of Agriculture*, vol. IV, 1832–4, 791–3.
Miller, Philip, *The Gardener's Dictionary*, 9th edn, London 1735.
Milne, William J., *Reminiscences of an old boy: being autobiographic sketches of Scottish rural life from 1832 to 1856*, Forfar 1901.
Mudie, Robert, *Spring, or, the causes, appearances, and effects of the seasonal renovations of nature in all climates*, London 1837.
Neill, Patrick, *On Scottish Gardens and Orchards. Drawn up by desire of the Board of Agriculture*, Edinburgh 1813.
Nevay, John, *The Peasant: a poem in nine cantos: and other poems*, Edinburgh 1834.
Mr Nicol, 'Directions for the culture of cottage-gardens', in Sir John Sinclair, *General Report of the Agricultural State and Political Circumstances of Scotland*, Edinburgh 1814, vol. II, chapter IX, appendix, article C, 426–7.
Oliver, T. 'The diet of toil', *The Lancet*, 29 June 1895, 1629–35.
Paterson, C. A., 'The housing of farm servants', *The Scottish Farm Servant*, 1st May 1913.
Paterson, Nathaniel, *The Manse Garden*, Glasgow 1836.
Pavy, F. W., *A Treatise on Food and Dietetics, Physiologically and Therapeutically Considered*, London 1874.
Pennant, Thomas, *A Tour in Scotland 1769* [1776], Perth 1979.
Philip, Adam, *The Parish of Longforgan: a sketch of its church and people*, Edinburgh 1895.

Reid, John, *The Scots Gard'ner*, Edinburgh 1683.

Rennie, Robert, 'Plan of an inland village', *THASS*, 1803, 250–66.

Robertson, George, *General View of the Agriculture of the County of Kincardine*, Edinburgh 1794.

—, *General View of the Agriculture of the County of Midlothian*, Edinburgh 1795.

—, *Rural Recollections; or, the progress of improvement in agriculture and rural affairs*, Irvine 1829.

Robertson, James, *General View of the Agriculture of the County of Perth*, Edinburgh 1813.

Scarfe, Norman (ed.), *To the Highlands in 1786: the inquisitive journey of a young French aristocrat*, Woodbridge 2001.

Scobell, Thomas, 'On the benefits of field gardens for the labouring poor', reported in Miscellaneous Notices, *The Quarterly Journal of Agriculture*, vol. IV, 1832–4, 99.

Seebohm Rowntree, B. and M. Kendall, *How the Labourer Lives, a study of the rural labour problem*, London 1913.

Shirra Gibb, Robert, *A Farmer's Fifty Years in Lauderdale*, Edinburgh 1927.

Sinclair, Sir John (ed.), *The Statistical Account of Scotland*, Edinburgh, 1791–9, available at https://stataccscot.edina.ac.uk/static/statacc/dist/home.

—, *An Account of the System of Husbandry Adopted in the More Improved Districts of Scotland, with Some Observations on the Improvements of which they are Susceptible*, Edinburgh 1812.

—, *General Report of the Agricultural State and Political Circumstances of Scotland*, Edinburgh 1814.

Skene Keith, George, *General View of the Agriculture of the County of Aberdeen*, Edinburgh 1811.

Skirving, Robert Scot, *Landlords and Labourers*, Edinburgh 1862.

Smiles, Samuel, *The Life of a Scotch Naturalist*, London 1905.

Smith, Edward, *Foods*, 2nd edn, London 1873.

Smith, George, 'Essay on the construction of cottages suited for the dwellings of the labouring classes, and adapted to the climate of Scotland', *THASS*, 1835, 205–16.

The Society for the Sons and Daughters of the Clergy, *New Statistical Account of Scotland*, Edinburgh 1834–1845. available at https://stataccscot.edina.ac.uk/static/statacc/dist/home.

Somerville, Alexander, *The Autobiography of a Working Man, by one who has 'whistled at the plough'*, London 1848.

—, *The Whistler at the Plough*, Manchester 1852.

Somerville, Robert, *General View of the Agriculture of the County of East Lothian*, Edinburgh 1813.

Somerville, Thomas, *My Own Life and Times 1741–1814*, Edinburgh 1861.

Southey, Robert, *Journal of a Tour in Scotland in 1819*, ed. C. H. Herford, Edinburgh 1929.

Stephen, Henry, *The Book of the Farm*, vol. 2, Edinburgh 1847.

Still, Peter, 'A Cottar's Sunday', in *The Cottar's Sunday and Other Poems, chiefly in the Scottish dialect*, Aberdeen 1844.

Substance of Sir John Sinclair's Addrese to the Board of Agriculture, *Scots Magazine*, 1st August 1795, 22–5.

Sweet, Alexander, *Villa and Cottage Gardening Specially adapted for Scotland, Northern England and Ireland*, 10th edn, London 1921.

Tait, James, 'Physiological distinctions in the condition of the Scotch peasantry', *THASS*, 1883, 1–20.

Taylor, George, *From Kelso to Kalamazoo: the life and times of George Taylor 1803–1891*, ed. Margaret Jeary and Mark A. Mulhearn, Edinburgh 2009.

Thompson, E. P., *The Making of the English Working Class*, London 1963.

Thompson, Flora, *Lark Rise to Candleford* [1945], Harmondsworth 1973.

Thomson, Alexander, 'On the settlement of crofters', *THASS*, 1837, 379–81.

Thomson, James, *Poems in the Scottish Dialect*, Edinburgh 1801.

Thomson, John, *General View of the Agriculture of the County of Fife*, Edinburgh 1800.

Trotter, James, *General View of the Agriculture of the County of West Lothian*, Edinburgh 1811.

Tully, John, *A half-century glance backwards: a paper read in Fochabers Mutual Improvement Association on February 12, 1930*, Banff 1930.

'Verus', 'On field gardening, and on the gardens of farm servants in Scotland', *GM*, 1826, 259–61.

Weathers, John, *A Practical Guide to School, Cottage and Allotment Gardening*, London 1908.

Wight, Andrew, *The Present State of Husbandry in Scotland. Extracted from Reports made to the Commissioners of the Annexed Estates*, 2 vols, Edinburgh 1778.

Young, Arthur, *Observations on the present state of the waste lands of Great Britain*, London 1773.

Younger, John, *Autobiography*, Kelso 1881.

W. J. I., 'Biographical sketch of Thomson the poet of Kenleith with a desultory description of his Parnassus', *Scots Magazine*, vol. 75, 1813, 645–8.

GOVERNMENT REPORTS

Board of Agriculture, *Report of the Committee of the Board of Agriculture appointed to extract information from the County Reports and other Authorities concerning the Culture and Use of Potatoes*, 1795.

Board of Agriculture and Fisheries, *Report on the Decline in the Agricultural Population of Great Britain 1881–1906*, London 1906.

Board of Agriculture for Scotland, *Agricultural Statistics 1913 vol. II pt 1.*

Factories Inquiry Commission, 1st Report, 1833.

Report from the Select Committee on Small Holdings, 1889.

Report of the Commissioner Appointed under the Provisions of the Act 5 & 6 Vict. c.99, to inquire into the operation of that Act and into the state of the population in the mining districts, 1844, by H. S. Tremenheere.

Report of HM Commissioners for Inquiring into the Housing of the Working Classes, 2nd Report: Scotland 1884–5.

Report on the food of the poorer labouring classes in England by Dr Edward Smith, in *6th Report of the Medical Officer of the Privy Council*, 1863.

Return of Allotments Detached from and Attached to Cottages, in England and Wales, and Scotland, 1886.

Royal Commission on the Employment of Children, Young Persons and Women in Agriculture, 4th Report, 1867.

Royal Commission on Labour: the agricultural labourer, vol. III Scotland pt I, Reports by H. Rutherfurd and G. R. Gillespie, 1893.

Royal Commission on Labour: the agricultural labourer, vol. III Scotland pt II, Reports by R. Hunter Pringle and Edward Wilkinson, 1893.

Royal Commission on Labour, Digest of Evidence vol. III Mining, Iron, Engineering and Hardware, 1893.

Royal Commission on Scientific Instruction and the Advancement of Science, Minutes of Evidence, vol. II, 1875.

Royal Commission on the Housing of the Industrial Population of Scotland, Rural and Urban, 1917–18.

OTHER REPORTS

1st Annual Report of the Association for Promoting the Improvement of the Dwellings and Domestic Condition of Agricultural Labourers in Scotland, Edinburgh 1855.

4th Annual Report of the Association for Promoting the Improvement of the Dwellings and Domestic Condition of Agricultural Labourers in Scotland, Edinburgh 1858.

Annual report by the County Sanitary Inspector Inverness-shire (Scotland), Inverness 1892.

'Hints by Lord Kames to the Commissioners of the Annexed Estates 1765', in Andrew Wight, *The Present State of Husbandry in Scotland. Extracted from Reports made to the Commissioners of the Annexed Estates*, vol. I, Edinburgh 1778.

NEWSPAPERS AND PERIODICALS

Aberdeen Journal, later the Aberdeen Press and Journal

Banffshire Journal

Berwick Advertiser

Berwickshire News

Blairgowrie Advertiser

Caledonian Mercury

Chambers Edinburgh Journal

Dalkeith Advertiser
Dollar Magazine
Dundee Courier
Dundee, Perth, Forfar and Fife People's Journal
Dunfermline Press
The Elgin Courant and Morayshire Advertiser
Fife Free Press
Fife Herald
Fifeshire Journal
Gardener's Magazine
Gardeners' Chronicle
Kelso Chronicle
Montrose, Arbroath and Brechin Review
North British Agriculturist
Perthshire Advertiser
Perthshire Courier
Scots Magazine
The Cheap Magazine
The Scotsman

SECONDARY BOOKS AND ARTICLES

Allen, David E. and Gabrielle Hatfield, *Medicinal Plants in Folk Tradition: an ethnobotany of Britain & Ireland*, Cambridge 2004.

Anderson, R. D., *Education and the Scottish People 1750–1918*, Oxford 1995.

Barnett, John, 'The English labourer revisited. Self-provisioning and the commercializing of diet', in M. Hietala and T.Vahkatari (eds), *The Landscape of Food*, Helsinki 2003.

Barrell, John, *The Dark Side of the Landscape: the rural poor in English painting 1730–1840*, Cambridge 1980.

Bartram, Thomas, *Bartram's Encyclopaedia of Herbal Medicine*, London 1998.

Bell, Jonathan and Mervyn Watson, *Rooted in the Soil: A History of Cottage Gardens and Allotments in Ireland since 1750*, Dublin 2012.

Brock, Jeannette M. 'The importance of emigration in Scottish regional population movement 1861–1911', in T. M. Devine (ed.), *Scottish Emigration and Scottish Society*, Edinburgh 1992.

Brown, Jane, *The Pursuit of Paradise*, London 1999.

Brown, Marilyn, *Scotland's Lost Gardens: from the Garden of Eden to the Stewart palaces*, Edinburgh 2012.

Buchan, David (ed.), *Folk Tradition and Folk Medicine in Scotland, the writings of David Rorie*, Edinburgh 1994.

Burchardt, Jeremy, 'Land and the Laborer: Potato Grounds and Allotments in Nineteenth-Century Southern England', *Agricultural History*, vol. 74, No. 3, 2000, 667–84.

—, *The Allotment Movement in England, 1793–1873*, Woodbridge 2002.

Burnett, John, 'The English village labourer revisited. Self-provisioning and the commercializing of diet', in M. Hietala and T. Vahkatari (eds), *The Landscape of Food*, Helsinki 2003.

Cameron, David Kerr, *The Ballad and the Plough*, London 1979.

Carpenter, Kenneth J. 'A short history of nutritional science: Part 1 (1785–1885)', *Journal of Nutrition*, vol. 133, 2011, 638–45.

Carter, Ian, 'The MI movement in NE Scotland in the 19th century', *Aberdeen University Review*, vol. XLVI, 4, no. 156, 1976, 383–400.

—, *Farm Life in Northeast Scotland 1840–1914: the poor man's country*, Edinburgh 1979.

Cheape, Hugh and Gavin Sprott, *Angus Country Life*, Edinburgh 1980.

Cooke, A. J., *Stanley – from Arkwright Village to Commuter Suburb: 1784–2003*, Perth 2003.

—, 'The early development of Stanley 1784–1823', in A. J. Cooke (ed.), *Stanley – Its History and Development*, Dundee 1977.

Cox, E. H. M., *A History of Gardening in Scotland*, London 1935.

Cox, Kenneth, *Scotland for Gardeners*, Edinburgh 2009.

Davidson, D. A. and T. C. Smout, 'Soil change in Scotland: the legacy of past land improvement processes', in A. G. Taylor, J. E. Gordon and M. B. Usher (eds), *Soils, Sustainability and the Natural Heritage*, Edinburgh 1996.

Davison, Fiona, *The Hidden Horticulturists: the untold story of the men who shaped Britain's gardens*, London 2019.

Dempster Hewitt, Mary, 'Robert Dempster: a forgotten activist', *Scottish Local History*, 88, 2014, 3–8.

Devine, T. M., 'Social stability in the Eastern Lowlands of Scotland during the agricultural revolution 1780–1840', Lairds and Improvement in the Scotland of the Enlightenment, 9th Scottish Historical Conference, Edinburgh, September 1978.

—, 'Scottish farm service in the agricultural revolution', in T. M. Devine (ed.), *Farm Servants and Labour in Lowland Scotland 1770–1914*, Edinburgh 1984.

—, 'Scottish farm labour in the era of agricultural depression 1875–1900', in T. M. Devine (ed.), *Farm Servants and Labour in Lowland Scotland 1770–1914*, Edinburgh 1984.

—, 'Social responses to agrarian "improvement": the Highland and Lowland clearances in Scotland, 1500–1850', in R. A. Houston and I. D. Whyte (eds), *Scottish Society 1500–1800*, Cambridge 1989.

—, *The Transformation of Rural Scotland: social change and the agrarian economy 1660–1815*, Edinburgh 1994.

—, *The Great Highland Famine*, Edinburgh 2004.

—, *The Scottish Clearances: a history of the dispossessed*, London 2018.

Dixon Hunt, John, 'Introduction', in J. Dixon Hunt and J. Wolske-Bulmahn (eds), *The Vernacular Garden – Dumbarton Oaks Colloquium Series in the History of Landscape Architecture*, Washington 1993.

Donald, Donna M., *Old Strichen, New Leeds and New Pitsligo*, Catrine 2011.

Dorward, David, *The Sidlaw Hills*, Balgavies 2004.

Drummond, J. C. and Anne Wilbraham, *The Englishman's Food: a history of five centuries of English diet*, London 1957.

Duncan, Robert, *The Mineworkers*, Edinburgh 2005.

Fenton, Alexander, *The Food of the Scots*, vol. 5 of *Scottish Life and Society, A Compendium of Scottish Ethnology*, Edinburgh 2007.

—, *The Hearth in Scotland*, Scottish Vernacular Buildings Working Group, Edinburgh 1981.

— and Bruce Walker, *The Rural Architecture of Scotland*, Edinburgh 1981.

Festing, S., *Gertrude Jekyll*, London 1993.

Fitzgibbon, Theodora, *A Taste of Scotland*, London 1970.

Flinn, M. W., 'Introduction', in Edwin Chadwick, *Report into the Sanitary Condition of the Labouring Population of Great Britain* [1842], ed M. W. Flinn, Edinburgh 1965.

— (ed.), *Scottish Population History*, Cambridge 1977.

Floud, Roderick, *An Economic History of the English Garden*, London 2019.

—, Kenneth Wachter and Annabel Gregory, *Height, Health and History. Nutritional Status in the United Kingdom, 1750–1980*, Cambridge 1990.

Gaskell, S. Martin, *Colliery Housing: a study in Midlothian*, Edinburgh 2011.

Gauldie, Enid, *The Scottish Country Miller 1700–1900*, Edinburgh 1981.

Genders, Roy, *The Cottage Garden*, London 1969.

Gibson, A. J. S. and T. C. Smout, 'Scottish food and Scottish history 1500–1800', in R. A. Houston and I. D. Whyte (eds), *Scottish Society 1500–1800*, Cambridge 1989.

—, *Prices, Food and Wages in Scotland 1550–1780*, Cambridge 1995.

Goddard, Nicholas, '19th century recycling. The Victorians and the agricultural utilisation of sewage', *History Today*, 31, June 1981, 32–6.

Gray, Malcolm, 'Farm workers in North-East Scotland', in T. M. Devine (ed.), *Farm Servants and Labour in Lowland Scotland 1770–1914*, Edinburgh 1984.

Greaves, P., 'Regional differences in the mid-Victorian diet and their impact on health', *Journal of the Royal Society of Medicine Open*, vol. 9, no. 3, 8th March 2018, 1–6.

Grigson, Jane, *English Food*, rev. edn, London 1992.

Haldane, Elizabeth S., *Scots Gardens in Old Times*, London 1934.

Holman, Bob, *Behind the Diamond Panes: the story of a Fife mining community*, Cowdenbeath 1952.

Houston, Rab, '"Frequent flitting": geographical mobility and social structure in mid-19th century Greenlaw', *Scottish Studies*, vol. 27, 1983, 27–42.

Hyams, Edward, *English Cottage Gardens*, London 1970.

Jackson, John Brinckerhoff, 'The past and present of the vernacular garden', in J. Dixon Hunt and J. Wolske-Bulmahn (eds), *The Vernacular Garden: 14th Dumbarton Oaks Colloquium Series on the History of Landscape Architecture*, revised papers, Washington 1993.

Jamieson, Lynn and Claire Toynbee, *Country Bairns: growing up, 1900–1930*, Edinburgh 1992.

Jones, Richard, 'Why manure matters', in Richard Jones (ed.), *Manure Matters – Historical, Archaeological and Ethnographic Perspectives*, Farnham 2012.

—, 'Understanding medieval manure', in Richard Jones (ed.), *Manure Matters – Historical, Archaeological and Ethnographic Perspectives*, Farnham 2012.

Kowalska-Lewicka, Anna, 'The potato and the Polish kitchen', in A. Fenton and Eszter Kisban (eds), *Food in Change – Eating Habits from the Middle Ages to the Present Day*, Edinburgh 1986.

Lamb, H. H., *Climate, History and the Modern World*, London 1995.

Laudan, Rachel, *Cuisine and Empire: cooking in world history*, Berkeley 2014.

Levitt, Ian and T. C. Smout, 'Farm workers' incomes in 1843', in T. M. Devine (ed.), *Farm Servants and Labour in Lowland Scotland 1770–1914*, Edinburgh 1984.

Lockhart, D. G., *Scottish Planned Villages*, Edinburgh 2012.

—, 'Planned villages in north east Scotland, 1750–1860', in J. Frew and D. Jones (eds), *The New Town Phenomenon: The Second Generation*, St Andrews Studies in the History of Scottish Architecture and Design IV, St Andrews 2000.

MacDougall, Iain, *Bondagers – Eight Scots Women Farm Workers*, East Linton 2000.

Marshall, David, Annie S. Anderson, Mike Lean and Ann Foster, 'Healthy eating: fruit and vegetables in Scotland', *British Food Journal*, 1st August 1994, 18–24.

McEwen, Ron, 'The northern lads: the migration of Scottish gardeners with especial reference to the Royal Botanic Garden, Kew', *Sibbaldia: The Journal of Botanic Garden Horticulture*, No. 11, 2013, 109–23.

McKean, Charles, *The Scottish Chateau: the country house of Renaissance Scotland*, Stroud 2001.

McNeill, F. Marian, *The Scots Kitchen*, new edn, Edinburgh 2010.

Millman, R. N., *The Making of the Scottish Landscape*, London 1975.

Mintz, Sidney W., *Sweetness and Power: the place of sugar in modern history*, New York 1985.

Mitchison, Rosalind, *Life in Scotland*, London 1978.

Mollison, Jenny, Judy Wilkinson and Rona Wilkinson, *Raising Spirits: allotments, well-being and community*, Edinburgh 2014.

Morgan, J. and A. Richards, *Paradise out of a Common Field: the pleasure and plenty of a Victorian garden*, London 1990.

Morrison, John, *Painting Labour in Scotland and Europe, 1850–1900*, Abingdon-on-Thames 2014.

Murray, Norman, *The Scottish hand loom weavers, 1790–1850: a social history*, Edinburgh 1978.

Nash, Andrew, 'The kailyard: problem or illusion?' in Susan Manning, Ian Brown, Thomas Owen Clancy and Murray Pittock (eds), *The Edinburgh History of Scottish Literature*, Edinburgh 2007.

Ord, John, *Ord's Bothy Songs and Ballads of Aberdeen, Banff and Moray, Angus and the Mearns* [1930], Edinburgh 1995.

Orr, Alistair, 'Farm servants and farm labour in the Forth valley and South-east Lowlands', in T. M. Devine (ed.), *Farm Servants and Labour in Lowland Scotland 1770–1914*, Edinburgh 1984.

Pears, Ben, 'The formation of anthropogenic soils across three marginal landscapes on Fair Isle and in the Netherlands and Ireland', in Richard Jones (ed.), *Manure Matters – Historical, Archaeological and Ethnographic Perspectives*, Farnham 2012.

Quest-Ritson, Charles, *The English Garden: A Social History*, London 2001.

Readman, Paul, *Land and Nation in England – patriotism, national identity and the politics of land, 1880–1914*, Woodbridge 2008.

Rice, Catherine, *'All their Good Friends and Neighbours': the story of a vanished hamlet in Angus*, Dundee 2014.

Riggs, Paul, 'The standard of living in Scotland', in J. Komlos (ed.), *Stature, Living Standards and Economic Development*, Chicago 1994.

Robertson, Forbes W., 'Pears in old Scottish orchards', *Review of Scottish Culture*, vol. 22, 2010, 1–17.

Robson, M., 'The Border farm worker', in T. M. Devine (ed.), *Farm Servants and Labour in Lowland Scotland 1770–1914*, Edinburgh 1984.

Rose, Jonathan, *The Intellectual Life of the British Working Classes*, 2nd edn, New Haven 2010.

Salaman, R. N., *History and Social Influence of the Potato*, rev. edn, Cambridge 1985.

Scarborough, Paul, R. D. Morgan, P. Webster et al., 'Differences in coronary heart disease, stroke and cancer mortality rates between England, Wales, Scotland and Northern Ireland: the role of diet and nutrition', *BMJ Open*1:e000263. doi: 10.1136/bmjopen-2011–000263.

Schneider, N. and W. Eugster, 'Climatic impacts of historical wetland drainage in Switzerland', *Climatic Change*, vol. 80, 2007, 301–21.

Scourse, Nicolette, *The Victorians and their Flowers*, London 1983.

Shaw, John, 'Manuring and fertilising the Lowlands', in S. Foster and T. C. Smout (eds), *The History of Soils*, Aberdeen 1994.

Smith, Catherine, 'A grumphie in the sty: an archaeological view of pigs in Scotland, from their earliest domestication to the agricultural revolution', *Proceedings of the Society of Antiquaries Scotland*, vol. 130, 2002, 705–24.

Smout, T. C., *A History of the Scottish People 1560–1830*, London 1969.

—, *A Century of the Scottish People 1830–1950*, London 1986.

—, *Nature Contested: environmental history in Scotland and Northern England since 1600*, Edinburgh 2000.

Steven, Maisie, *The Good Scots Diet*, Glendaruel 2003.

Stocks, Christopher, *Forgotten Fruits: a guide to Britain's traditional fruit and vegetables*, London 2008.

Symon, J. A., *Scottish Farming Past and Present*, Edinburgh 1959.

Thomas, Keith, *Man and the Natural World: Changing Attitudes in England 1500–1800*, London 1983.

Uglow, Jennifer S., *A Little History of British Gardening*, London 2004.

Watson, Roderick, *The Literature of Scotland: the Middle Ages to the 19th century*, London 2007.

Westmacott, Richard, *African-American Gardens and Yards*, Knoxville 1992.

Whittington, Graeme, 'The Little Ice Age and Scotland's weather', *Scottish Geographical Magazine*, vol. 101, no. 3, 1985, 174–8.

Whyte, Ian, *Agriculture and Society in 17th century Scotland*, Edinburgh 1979.

—, 'Population mobility in early modern Scotland', in R. A. Houston and I. D. Whyte (eds), *Scottish Society 1500–1800*, Cambridge 1989.

Wightman, Andy, *The Poor Had No Lawyers: who owned Scotland (and how they got it)*, Edinburgh 2010.

Willes, Margaret, *The Gardens of the British Working Class*, New Haven 2014.

Woodforde, John, *The Truth About Cottages* [1969], London 2007.

Woodward, Donald, '"Gooding the earth": manuring practices in Britain 1500–1800', in S. Foster and T. C. Smout (eds), *The history of soils and field systems*, Aberdeen 1994.

WEBSITES

Dictionary of the Scots Language, http://www.dsl.ac.uk

The Friends of Forfar Botanists, http://www.forfarbotanists.org/georgedon.html

Future Museum, http://www.futuremuseum.co.uk/collections/people/key-people/burns/gavin-hamilton.aspx

Groome's Ordnance Gazetteer of Scotland 1882, http://gazetteerofscotland.org.uk

https://www.gardenvisit.com/book/gardening__tours_by_jc_loudon_1831-1842/manchester_chester_liverpool_and_scotland_in_the_summer_of_1831

Orange Pippin website, https://www.orangepippin.com/varieties/apples/ribston-pippin

Public Health (Scotland) Act 1897, https://www.legislation.gov.uk/ukpga/Vict/60-61/38/contents/enacted

Scotland's People, https://www.scotlandspeople.gov.uk

Scottish Archive Network Weights and Measures Guide, https://www.scan.org.uk/measures/capacity.asp

Scottish government, https://www.gov.scot/publications/scottish-health-survey-2012-volume-1-main-report/pages/16/6

Scottish Mining, http://www.scottishmining.co.uk/461.html

Scottish Post Office Directories, NLS e-resources, https://digital.nls.uk/directories

CONFERENCE PAPERS

Bruneel, Dieter, 'Rural property offences and the changing nature of landless households in 19th century Flanders', Rural History Conference Leuven, September 2017.

Hallgren, Karin, 'Vegetable growing by peasants in 18th century Sweden', Rural History Conference, Leuven, September 2017.

Miettinen, Riika, 'The livelihood strategies of the landless labourers in the early modern Swedish countryside', Rural History Conference, Leuven, September 2017.

Paping, Richard, 'The rise of landless households in the Dutch countryside ca 1600–1900', Rural History Conference, Leuven, September 2017.

UNPUBLISHED THESES

Burns, Catherine M., 'Industrial labour and radical movements in Scotland in the 1790s', MSc thesis, University of Strathclyde 1971.

Desilvey, Caitlin, '"When plotters meet": Edinburgh's Allotment Movement 1921–2001', MSc thesis, University of Edinburgh 2001.

Reid, Deborah, 'Unsung Heroines of Horticulture: Scottish Gardening Women, 1800–1930', PhD thesis, University of Edinburgh 2015.

Smith, Annette M., 'The Forfeited Estates Papers, 1745 – A study of the work of the commissioners for the Forfeited Annexed Estates, 1755–1784, with particular reference to their contribution to the development of communications in Scotland in the eighteenth century', PhD thesis, University of St Andrews 1974.

Wilkinson, Anne, 'The Development of gardening as a leisure activity in 19th century Britain and the establishment of horticultural periodicals', PhD thesis, Open University 2002.

INDEX

Note: Page numbers in *italic* refer to figures, those in **bold** to tables. Footnotes are indicated by a lower-case 'n' followed by the note number, e.g. 62 n.4.

Garden and Landscape History

Previously Published

Designs Upon the Land: Elite Landscapes of the Middle Ages
Oliver H. Creighton

Richard Woods (1715–1793): Master of the Pleasure Garden
Fiona Cowell

Uvedale Price (1747–1829): Decoding the Picturesque
Charles Watkins and Ben Cowell

Common Land in English Painting, 1700–1850
Ian Waites

Observations on Modern Gardening, *by Thomas Whately: An Eighteenth-Century Study of the English Landscape Garden*
Michael Symes

The Landscape Studies of Hayman Rooke (1723–1806): Antiquarianism, Archaeology and Natural History in the Eighteenth Century
Emily Sloan

Transhumance and the Making of Ireland's Uplands, 1550–1900
Eugene Costello

The Foldcourse and East Anglian Agriculture and Landscape, 1100–1900
John Belcher

Erasmus Darwin's Gardens: Medicine, Agriculture and the Sciences in the Eighteenth Century
Paul. A. Elliott

Rediscovering Lost Landscapes: Topographical Art in north-west Italy, 1800–1920
Pietro Piana, Charles Watkins and Ross Balzaretti